Planetary mapping

Cambridge Planetary Science Series

EDITORS: W. I. Axford, G. E. Hunt, and R. Greeley

PLANETARY MAPPING

Edited by

RONALD GREELEY
Arizona State University

RAYMOND M. BATSON
United States Geological Survey

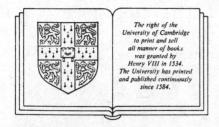

The right of the
University of Cambridge
to print and sell
all manner of books
was granted by
Henry VIII in 1534.
The University has printed
and published continuously
since 1584.

Cambridge University Press

Cambridge

New York   Port Chester   Melbourne   Sydney

CAMBRIDGE UNIVERSITY PRESS
Cambridge, New York, Melbourne, Madrid, Cape Town, Singapore, São Paulo

Cambridge University Press
The Edinburgh Building, Cambridge CB2 2RU, UK

Published in the United States of America by Cambridge University Press, New York

www.cambridge.org
Information on this title: www.cambridge.org/9780521307741

First published 1990
This digitally printed first paperback version 2006

A catalogue record for this publication is available from the British Library

Library of Congress Cataloguing in Publication data
Planetary mapping / edited by Ronald Greeley, Raymond Batson.
    p.   cm. – (Cambridge planetary science series; 6)
    Includes bibliographical references.
    ISBN 0-521-30774-0
    1. Planets – Maps.   2. Satellites – Maps.   I. Greeley, Ronald.
II. Batson, Raymond M.   III. Series.
QB605.P58   1990
912.99 – dc20   90-1418
                                                                            CIP

ISBN-13 978-0-521-30774-1 hardback
ISBN-10 0-521-30774-0 hardback

ISBN-13 978-0-521-03373-2 paperback
ISBN-10 0-521-03373-X paperback

Contents

CONTENTS

Preface

Although mapmaking dates from earliest recorded history, the systematic mapping of the planets and satellites is mostly a product of the space age. With the successful flyby of the planet Neptune by Voyager in 1989, nearly all of the major objects in the Solar System have been seen via spacecraft and at least partly mapped. This book outlines the methods and techniques used in making maps of planetary objects and was prepared for all readers interested in Solar System exploration and mapmaking. The chapters are written by mapmakers who developed these methods and are leaders in their respective disciplines. We are grateful for the care and consideration of the authors in preparing the chapters and hope that you will find their contributions as useful as we have in understanding the Solar System.

We acknowledge Cynthia Greeley for preparing the index to this book. We also wish to recognize that most of the planetary mapping conducted in the United States has been supported by the National Aeronautics and Space Administration, primarily through the Planetary Geosciences Program under the direction of Stephen Dwornik and Joseph Boyce.

Ronald Greeley and Raymond Batson

Contributors

Raymond M. Batson
United States Geological Survey
Flagstaff, Arizona

Merton E. Davies
RAND Corporation
Santa Monica, California

Frederick J. Doyle
United States Geological Survey
Reston, Virginia

Ronald Greeley
Department of Geology
Arizona State University
Tempe, Arizona

Jay L. Inge
United States Geological Survey
Flagstaff, Arizona

Harold Masursky
United States Geological Survey
Flagstaff, Arizona

Mary Emma Strobell
United States Geological Survey
Flagstaff, Arizona

Ewen A. Whitaker
Lunar Planetary Laboratory
University of Arizona
Tucson, Arizona

Don E. Wilhelms
United States Geological Survey
Menlo Park, California

Sherman S. C. Wu
United States Geological Survey
Flagstaff, Arizona

Planetary mapping

1

Introduction

RONALD GREELEY AND
RAYMOND M. BATSON

In the past two decades the exploration of much of the Solar System has become a reality. More than one hundred missions have been flown to the Moon and planets by the United States and the Soviet Union. Many of these missions have carried imaging systems that, collectively, have returned an incredible wealth of information on the shape and surface characteristics of planetary objects.

Throughout history, maps and charts have played an integral role in the exploration of Earth. Their importance holds true for Solar System exploration as well. Maps of the planets are needed by planners of spaceflights to design missions, including the selection of safe and scientifically fruitful landing sites, and are the framework for recording measurements from a wide variety of spacecraft instruments. During the data analysis phase following the completion of mission operations, maps and charts provide the basis for understanding local, regional, and global characteristics of planets and satellites.

The making of planetary maps has required the development of new methods and techniques. Many of the basic principles derived from the mapping of Earth must be reconsidered in the mapping of other planets. For example, the traditional datum on Earth is sea level, but what does one select as the datum on planets without oceans? Most topographic data on Earth are derived photogrammetrically from conventional aerial photographs. How can these techniques be applied to a planet completely covered in perpetual clouds, such as Venus, where only radar images can be obtained? How can the geological evolution of a planet be determined when its surface has not been visited and when photographs are the primary data that are available? These problems have been addressed by planetary mapmakers since the early 1960s, and as is true for much of Solar System exploration, some of the answers have direct application to Earth.

As reviewed in Chapter 2, mapping our closest planetary neighbor, the Moon, began in the seventeenth century with Galileo. Rapid advances in the quality of telescopes led to the preparation of detailed maps of the Moon and the establishment of formal procedures for naming surface features (see Chapter 4). Simple maps were also produced for Mars and Mercury based on telescopic observations. But by the early 1900s astronomers had determined the orbits, rotations, sizes, masses, and other whole-body properties for the planets and many satellites almost to their satisfaction and had turned their attention to stellar and galactic problems that they regarded as more significant. Consequently, the study of the Moon and planets was in a state of neglect when Sputnik 1 initiated the Space Age in October 1957.

During the period of neglect prior to the launch of Sputnik 1, a few pioneers, such as Ernst Opik, Ralph B. Baldwin, Harold C. Urey, Gerard P. Kuiper, and Eugene M. Shoemaker, were quietly paving the way for what would become an intensive and productive program of lunar and planetary investigations, including mapmaking. Official U.S. interest in the solid bodies of the Solar System increased when the National Aeronautics and Space Administration (NASA) was established in October 1958 in response to the success of the Sputniks. In 1959 the race for the Moon began in earnest when the first spacecraft, the Soviet Lunas 2 and 3, reached the Moon, and when the United States initiated Project Ranger.

With the formation of NASA, major cartographic and geologic mapping programs were also initiated. Working with Gerard Kuiper and Zdenek Kopal at the University of Manchester, the U.S. Air Force Aeronautical Chart and Information Center (ACIC) began its fundamental program of lunar cartography, which has been continued and extended to the planets and outer satellites by the U.S. Geological Survey. The U.S. Army Corps of Engineers commissioned a lunar study by the Military Branch of the U.S. Geological Survey that included the first modern geologic map of the lunar surface. Shoemaker began his highly productive studies of impact and volcanic craters, geologically mapped a key area of the Moon, and launched a major effort of lunar geologic mapping that has similar programs for each newly photographed planet and satellite (see Chapter 7).

1.1. PLANETARY VERSUS TERRESTRIAL MAPPING

Through the centuries, maps of Earth's surface have been produced primarily by piecing together large-scale sketches and diagrams. "Control" networks were derived through extensive and laborious ground surveying. By the late nineteenth century, regional maps were produced in this fashion that were relatively accurate. With twentieth-century technology came the ability to obtain the so-called synoptic view. Photographs taken first from aircraft and later from Earth-orbiting satellites enabled the rapid production of accurate maps. When combined with well-established control networks, these maps have enabled surface features on Earth to be located precisely.

Planetary explorers, on the other hand, have had the global perspective from the beginning, and they have progressed from global, through regional, to local vantages. Solar System exploration – and the production of planetary maps – typically involves data reduction from a sequence of progressively more complex missions. The first stage in the exploration of a planet typically involves "flyby" reconnaissance missions, in which a spacecraft takes pictures and records other data as it passes a planet or planetary system. The Voyager mission through the outer Solar System is an example of a flyby mission in which careful planning allowed extensive data returned from both the major planets and from the moons that surround them (Figure 1.1). Rapidly rotating planets can be imaged completely as a spacecraft approaches and departs, but data resolution is extremely variable because different longitudes are viewed from different distances. The pictures of Rhea in Figure 1.2, for example, show typical ranges of image resolutions in flyby missions.

Later missions involve orbiting spacecraft, such as Mariner 9 and the Viking Orbiters to Mars. Orbiters enable systematic collection of data at a consistent range of resolutions. A planet rotating beneath the elliptical orbit of a spacecraft presents mission planners with a variety of data-gathering possibilities. For example, lighting geometry greatly influences the type of information returned by images. As shown in Figures 1.3 and 1.4, pictures taken when the Sun casts shadows reveal surface features such as small hills had fractures; in contrast, pictures taken when the Sun is high reveal the reflective properties of planetary surfaces.

Landings on planetary surfaces are made in the advanced stages of exploration. Although the primary objectives of such missions do not include mapping, very-large-scale maps are made to record data gathered around the landing sites and, if appropriate, to document sample collection sites. These missions also have important geodetic significance because the coordinates of the landing sites can be located precisely. Many techniques have been used for this purpose, including laser-ranging from observatories on Earth to corner reflectors left on the Moon by Apollo astronauts and analysis of radio-tracking data from the Viking Landers on Mars. When precise locations of landing sites can be positioned on mapping images, the measurements serve the very important function of tying a planetary control net to monumented bench marks on the ground (see Chapter 5).

The naming of features is as much a part of mapmaking as are the measuring and plotting of their locations. Without names, communication of ideas is impossible. The names applied by explorers on Earth often bear their provincial outlook. Ambiguities abound; settlers on different parts of the same river often know the river by different names. Invaders rename the territories of the invaded.

The tradition that the privilege of naming belongs to the discoverer resulted (on the Earth as on the planets) in hopeless ambiguities, redundancies, and inconsistencies. The International Astronomical Union (IAU) has therefore assumed control of the naming process. Its working groups are composed of planetary scientists from many nations, and although the process is often an emotional and politically contentious one, a system of nomenclature has emerged that is accepted by the international scientific community, as discussed in Chapter 4.

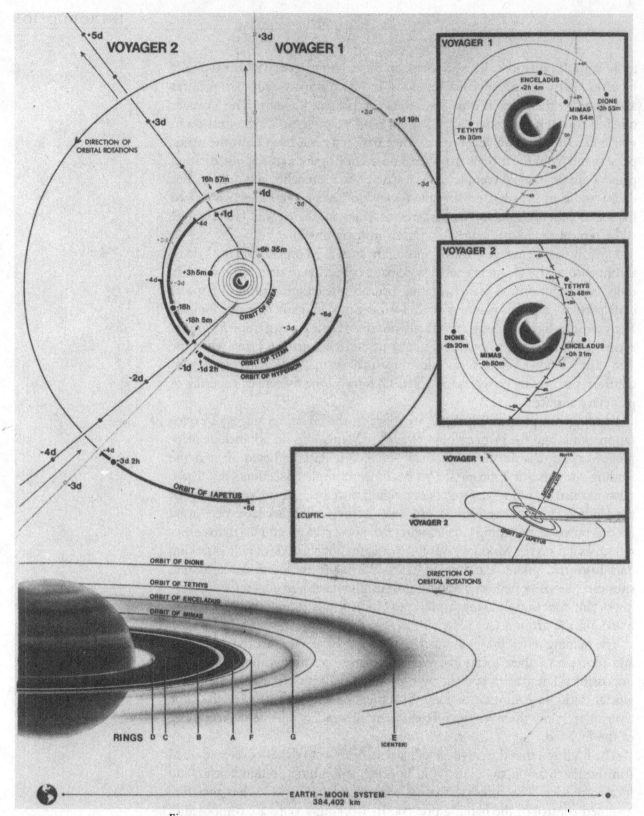

Figure 1.1
The flights of the Voyagers through the Saturnian system. Voyager 1 was deflected northward out of the ecliptic, whereas Voyager 2 was programmed to use the gravity of Saturn to propel it on to Uranus and eventually to Neptune. The Earth–Moon system is shown for scale.

1.2. DATA FOR MAKING PLANETARY MAPS

Monoscopic television images are the primary resource for mapping the planets. Techniques for making maps from aerial photographs of the Earth were developed many years ago, and a stereoscopic approach to mapping was used almost from the beginning. Terrestrial methods have been refined and modified for specialized use on the planets, but there are several important differences between terrestrial aerial photographs and digital television images returned by spacecraft.

Most aerial cameras have fields of view of 90 degrees or more across a frame, whereas spacecraft television images commonly have fields of view of less than 2 degrees, and in some cases are as small as one-tenth of a degree. Aerial-mapping cameras have minimal, almost unmeasurable, geometric distortions; spacecraft television systems tend to produce images that are too distorted for cartographic use until they have been modified through complex computer processing. The spatial resolution of an aerial image is limited by the grain size of the film emulsion, whereas the resolution of a television image is limited by the size of a digital picture element, or "pixel." The area of the film plane of an aerial camera is more than 500 times larger than the image plane of a spacecraft television camera; if television vidicon tubes and film could be compared directly in terms of sensitivity and resolution, film images would contain five hundred times as much spatial information as television images.

Most importantly, for applications to Earth, aerial film cameras and photogrammetric mapping instruments were designed as components of precision mapping systems to utilize the stereoscopic effect available in overlapping images taken from different points of perspective. Such image pairs are rare in planetary exploration, and even when they are available, their geometry is incompatible with most stereoscopic mapping instruments. These and other factors make the aerial film camera a far more effective mapping tool than spacecraft television systems. Except for some Apollo and Soviet missions, however, it has not been practical to use film cameras to obtain data for making extraterrestrial maps.

There are, however, some advantages in the use of digital television cameras rather than film systems. The television camera can record a much wider range of brightness values than film, and it can record them more precisely. Consequently, more subtle color changes can be recorded electronically than on film. In addition, the total number of images recovered from a mission is limited only by the power available on the spacecraft. From a mission like Voyager, which has lasted for more than a decade, tens of thousands of pictures can be acquired. Television images are returned to Earth at the speed of light, rather than at the speed of a spacecraft. Thus, Voyager images of the Saturnian system were recovered only 1.5 hours after they were taken, whereas it would have taken 2 or more years to send them back to Earth by a returning spacecraft!

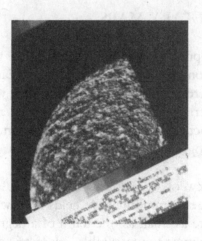

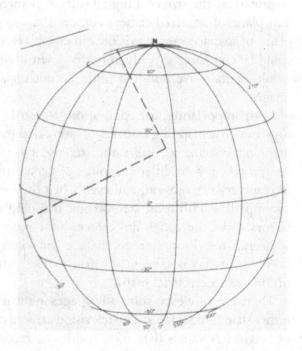

Figure 1.2
Voyager 1 pictures of Rhea, a satellite of Saturn. On the facing page, the image on the right (PICNO 115S1-002) was taken from a distance of 1,850,000 km; the one on the left (PICNO 155S1-001), from 720,000 km. The facing image (PICNO 0338S1 + 000) was taken from a distance of 121,000 km. The perspective grids below each picture were derived from the radio-tracked positions of the spacecraft at the time each exposure was made. Note that only albedo patterns are apparent on pictures taken from great distances; image resolution must be better than 2 or 3 km per pixel to show landforms.

The image plane of most spacecraft cameras is the surface of a television vidicon tube. A pattern of reference marks called a reseau is inscribed in the phosphor of the tube (Figure 1.5). The coordinates of these marks are measured prior to spacecraft launch so that distortion can be measured and the correct geometry can be reconstructed when the images are received from a spacecraft. Vidicon systems have an undesirable characteristic: The electronic beam used to scan an image from the vidicon tube is deflected slightly when

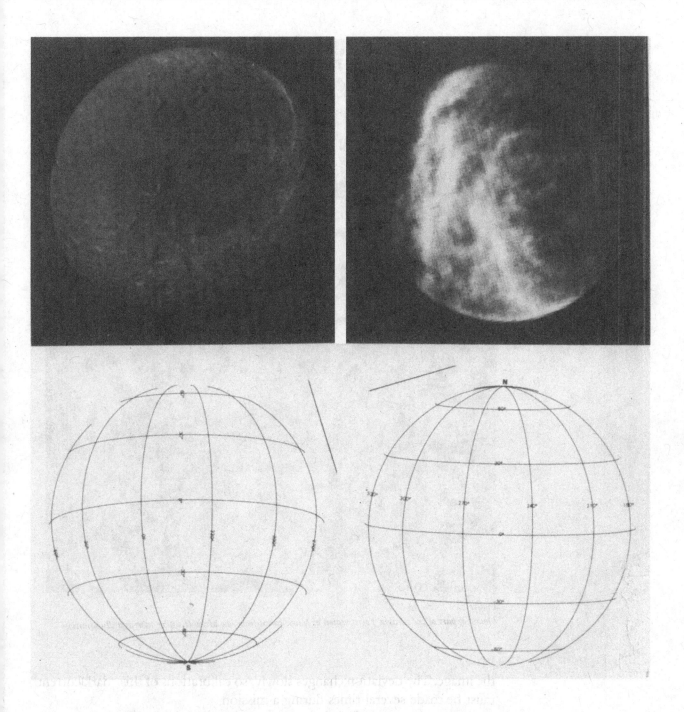

it encounters bright areas, so image geometry can never be precisely reconstructed.

Imaging tubes vary in sensitivity over the image area. Some of this variation is inherent in the manufacture of the tube and can be calibrated prior to spacecraft launch, but some is caused by temperature changes, strong radiation, and magnetic fields, such as those that surround Jupiter and Saturn. These effects cannot be predicted prior to the mission and must be calibrated in flight by taking pictures of black sky and measuring deviation from black in

Figure 1.3
Mosaic of part of the Arabia Terra region of Mars, taken from low altitude under morning illumination.

the image. The deviation changes slowly, so calibrations of this "dark current" must be made several times during a mission.

The vidicon television tube has been replaced on planetary imaging systems of the 1990s with a "charge coupled device," or CCD. The CCD consists of an array of tiny solid-state sensors, the location of each of which is precisely known, covering an entire image plane. This system, planned for its first planetary use on the Galileo mission to Jupiter, will be a boon to cartographers and photogrammetrists because many of the difficulties with the vidicon system will be eliminated, such as the beam-bending and reseau-fitting problems.

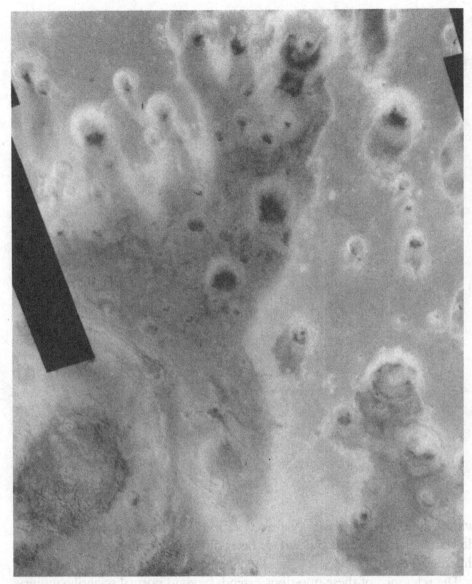

Figure 1.4
Mosaic of part of the Arabia Terra region taken from high altitude under noontime illumination.

1.3. TOPOGRAPHIC MAPS AND GEODETIC CONTROL OF PLANETARY SURFACES

Topographic measurements provide a third dimension to maps by showing the height of features above some reference elevation. This dimension enables more complete studies of surface morphology, structure, and local gravity. Without this additional dimension, many scientific questions are difficult or impossible to answer. For example, what is the volume of a river channel, and by implication, what is the volume of the material that was excavated to form the channel? How high are the mountains, and how deep are the basins

Figure 1.5
Image of black sky, taken by a
Voyager camera for radiometric
calibration. Note the pattern of
reseau marks, used for geometric
correction of the image.

with respect to local gravitational acceleration? Which way might liquids have flowed in the past? Has the surface been tilted so that they would flow in a different direction today?

The relevance of these questions extends beyond normal scientific inquiry; planners of future missions are vitally interested not only in the scientific questions, but in the operational ones as well. Can a remotely controlled vehicle climb a particular slope? Is a selected site sufficiently smooth for a spacecraft landing? Will the descent trajectory of the spacecraft intersect the surface at the desired landing site, or is there a mountain in the way?

Measuring topography on the planets is very different than it is on Earth. On Earth extensive oceans and shorelines, absent on planets explored to date, provide an obvious datum from which to measure elevations. Sea level is a surface that is linked to the shape of the Earth's gravitational field, incorporating a necessary geophysical element. Definition of a topographic datum on other planets is less direct. The gravity field of the Moon, for example, was defined by radio tracking of orbiting spacecraft, and a topographic datum based on this field was proposed. Similarly, the topographic datum on Mars was defined by its gravity field. The gravity field of Venus, obtained from the Pioneer Venus mission, indicates that a spherical figure is adequate for mapping.

In the early stages of planetary exploration, NASA has necessarily emphasized high-resolution global imaging of the planets. This mandates the use of long focal-length camera lenses, resulting in very narrow fields-of-view. In addition, images are taken under a wide variety of viewing geometries and resolutions. These conditions are antithetical to the systematic accumulation of data for topographic mapping. The topographer must therefore deal with a variety of unorthodox datasets, as discussed in Chapter 6.

Controls for planetary maps must also be found by indirect means; field surveys of the sort used on Earth are not possible. Planetary map controls are usually derived by methods of photogrammetric triangulation specially developed for use with spacecraft images. Computations are supported by Earth-based tracking of spacecraft and by astronomical ephimerides that give the relative positions of the spacecraft and the surface being mapped. Photogrammetric triangulation is a process by which the geodetic positions of images of selected surface features are computed from precise measurements of their position on two or more photographs or television pictures. Given a collection of overlapping pictures taken from many points of perspective around a planet, the latitude, longitude, and altitude of an array of specifically identified points (surface features such as craters) can be computed. These points are then used to control the positioning of all other features on a map.

1.4. SUMMARY

The generation of planetary maps has posed a wide variety of challenges to the scientific community and to cartographers. For the most part, these challenges have been met successfully through the development of innovative cartographic techniques.

In the chapters that follow, the key elements involved in making maps of planets and satellites are presented, problems and their solutions are discussed, and prospects for the future are outlined. Planetary mapping has contributed substantially to the understanding of planetary objects and will continue to play a key role in Solar System exploration.

2

History of planetary cartography

RAYMOND M. BATSON,
EWEN A. WHITAKER, AND
DON E. WILHELMS

2.1. LUNAR AND PLANETARY MAPPING FROM THE EARTH

2.1.1. *Lunar cartography and photography (1600–1967)*

The Moon is unique among Solar System bodies in the sense that it is the only one that is close enough that undisputed topographic details can be seen through the telescope. Topography is thrown into prominent relief near the terminator (i.e., the sunrise/sunset line), but relief vanishes in regions well removed from the terminator and also at the full phase, when the only markings visible are those caused by differences in reflectivity (albedo) of the surface materials. Many albedo markings (e.g., the crater rays) show no topographic expression under any illumination, whereas numerous topographic features, such as old, degraded craters, are totally invisible at the full phase.

THE PREPHOTOGRAPHIC ERA The terms "maps" or "charts" have been applied indiscriminately to many representations of the lunar surface, including photographs, drawings, and characterizations of the albedo, whether or not they included names or coordinate grid systems. Similarly, many collections of unannotated photographs are referred to as atlases.

Under these definitions, the earliest lunar maps date back to about 1600, with Gilbert's naked-eye full-Moon sketch with nomenclature (Figure 2.1); to 1610, with Galileo's telescopic sketches of phases; and to 1612, with Harriot's fairly detailed full-Moon drawing showing more than seventy letter and number designations (Figure 2.2). However, honors for compiling the first useful map go to Michael van Langren (Langrenus). His 34-cm-diameter lunar disk, published in 1645, combines the full-Moon albedo markings and topographic features from observations made at thirty different phases (Figure 2.3). The craters and mountains are portrayed as if everywhere illuminated by a fairly low, rising Sun. The image is oriented with the Moon's north and south poles exactly at the top and bottom, respectively, and with the mean

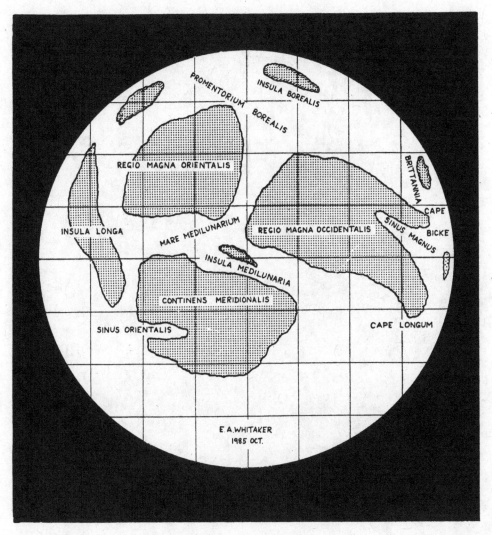

INSULA BOREALIS

PROMENTORIUM BOREALIS

REGIO MAGNA ORIENTALIS

BRITTANNIA

INSULA LONGA

MARE MEDILUNARIUM

REGIO MAGNA OCCIDENTALIS

CAPE

SINUS MAGNUS

BICKE

INSULA MEDILUNARIA

CONTINENS MERIDIONALIS

SINUS ORIENTALIS

CAPE LONGUM

E. A. WHITAKER
1985 OCT.

Figure 2.1
Reproduction of W. Gilbert's
naked-eye drawing of full Moon,
circa 1600 (based on MS in
British Library – original is not
suitable for reproduction).

center of face at the center of the disk (i.e., as seen at zero libration). Langrenus also initiated the current scheme of crater nomenclature in which the craters are identified by the names of famous people, but the names he applied to the maria did not survive him. He used 325 names in all.

The six years following the publication of Langrenus' map saw a plethora of lunar maps and images, but the only influential ones were those of Johann Höwelcke (Hevelius) in 1647 and of F. M. Grimaldi and G. B. Riccioli (Riccioli, 1651). Hevelius published his observations in a large tome, *Selenographia,* which includes forty images of the whole range of phases and three larger (28-cm-diameter) maps, all very artistically produced. One of these is an image of the full Moon; the second is similar in content to the Langrenus map but without any nomenclature (Figure 2.4), whereas the third is a rather peculiar representation of the topographic features using the then-current terrestrial convention of portraying mountain ranges as rows of African termite hills or hay piles as seen in perspective on a flat, level surface! On this third

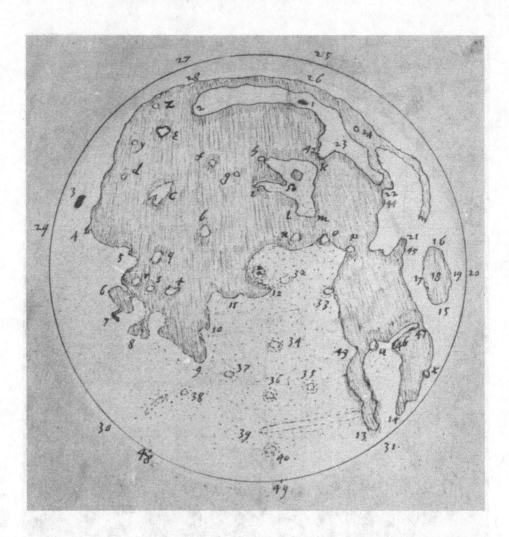

Figure 2.2
Thomas Harriot's full-Moon
map, circa 1612. (From MS in
possession of Earl of Egremont
and Leconfield.)

map, Hevelius includes nomenclature, which contains only ancient classical names from terrestrial geography. All three maps attempt to show the libration limits by using two displaced circles for the locations of the limb with respect to the surface features.

The Grimaldi map utilizes the best data from earlier maps, as corrected and augmented by visual observations, but its chief claim to fame is the nomenclature for craters and maria supplied by Riccioli (Figure 2.5). This follows Langrenus' scheme of using the names of famous personages, but Hevelius restricts them to philosophers, scientists, etc. It has stood the test of time (Riccioli's mare nomenclature is in use today) and forms the core of the present nomenclature (see Chapter 4).

What might have been the next milestone in lunar cartography failed to materialize because of restricted production and distribution. This was the "large" (54-cm-diameter) map produced under J. D. Cassini's direction in 1680. Although it is considerably more detailed than its predecessors, a result of the use of larger (and more unwieldy) telescopes, some of this detail is

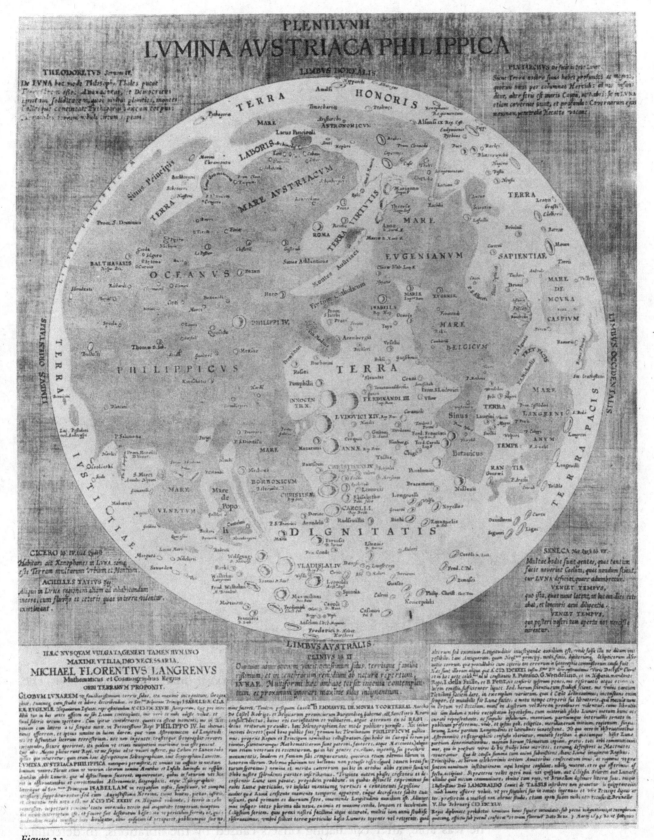

Figure 2.3
The first true map of the Moon, with albedo markings, topographic features and nomenclature, by Michael van
Langren (Langrenus), 1645. (From copy in Crawford Library, Royal Observatory, Edinburgh.)

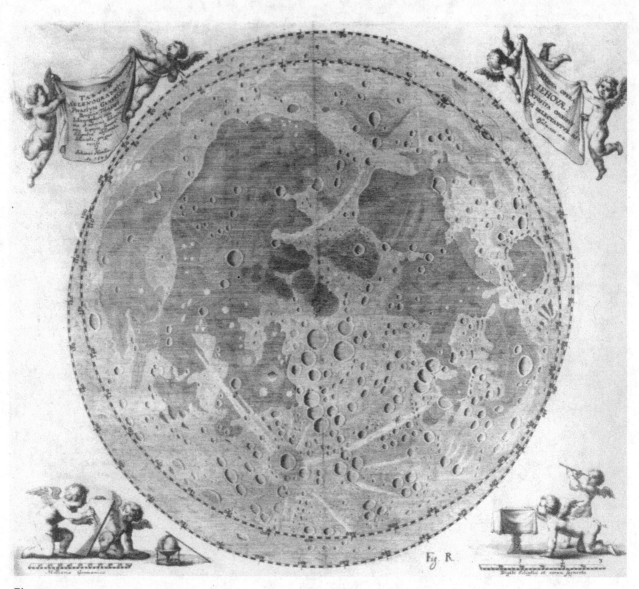

Figure 2.4
Johann Höwelcke's (Hevelius, 1647) map with albedo markings and topographic features. He also made two companion maps, one showing just the albedo markings (i.e., full Moon), the other providing his nomenclature.

either illusory or merely conventional, and positional accuracy is poorer. The map carries no nomenclature, and it is oriented with south "up" but rotated 23 degrees clockwise from vertical; it includes the main albedo markings and again uses the oblique solar illumination convention for portrayal of topography.

Lunar mapping did not materially advance until about seventy years later, in 1749, when Tobias Mayer produced the first map based on measured points (Lichtenberg, 1775). Using a telescope provided with a reticle of fine threads, he was able, after calculating offsets caused by the librations, to determine the absolute latitudes and longitudes of some ninety lunar features. He plotted these on an orthographic grid (i.e., a sphere viewed from infinity) with lines at 10-degree intervals, the remaining features being filled in from his set of

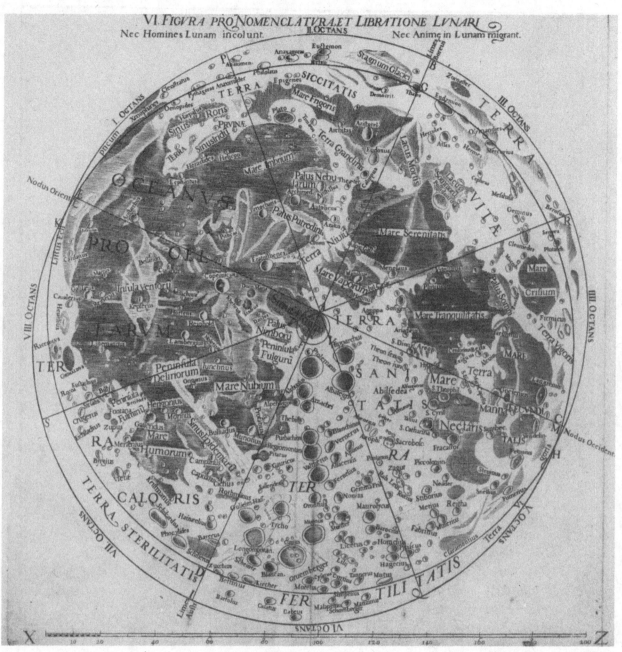

Figure 2.5
Grimaldi's map with Riccioli's nomenclature, which forms the basis of current nomenclature (Riccioli, 1651).

previously made drawings of various lunar areas. He also used the "morning illumination" technique for portrayal of the topographic features (Figure 2.6).

Toward the end of the eighteenth century, William Herschel had perfected methods for constructing reflecting telescopes that (compared with currently available refractors) had a large aperture and short focal length, thus allowing the observation of much finer lunar surface detail. Using such an instrument, together with an unwieldy one 18 inches in diameter, Johann Schröter made

Figure 2.6
Tobias Mayer's (1775) map, the first to be based on position measurements and to be provided with a coordinate net. Drawn circa 1749.

numerous drawings of many different areas of the disk, annotating them with the Riccioli nomenclature, additional names of his own, and many Roman and Greek letters. Although the shadows are portrayed as seen at the time of observation, crater rims are presented in a peculiar stylized form in which they resemble aerial views of rings of trees. All the drawings are oriented with south "up" (Figure 2.7).

The nineteenth century saw the production of four major lunar maps, three of which were compiled by German astronomers. The first was that of Wilhelm

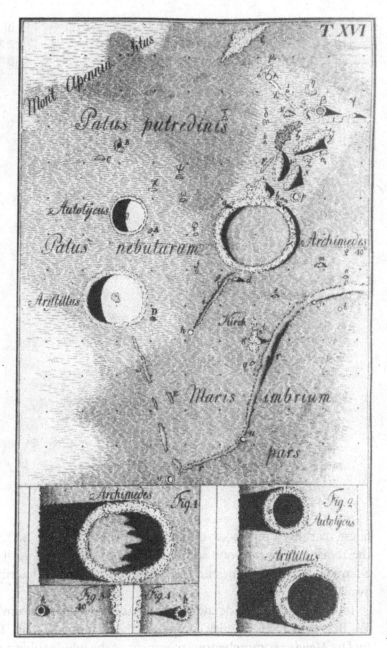

Figure 2.7
Example of typical drawings by
Johann Schröter (1791).

Lohrmann (1824), although only four of the twenty-five sections of the complete map were published during his lifetime. The scale is 97.5 cm to the lunar diameter, with south "up" and with the orthographic grid included at 5-degree intervals. Lohrmann introduced the art of hachuring to lunar cartography. Steepness of slopes is indicated by the thickness of the hachure lines, and albedo by the density of the background stippling, using a scale of 1 (darkest) to 10 (lightest). This map is remarkably detailed and accurate in view of the small (4.8-inch) telescope used; it was obviously drawn and engraved with meticulous care, and some users say that it is aesthetically the

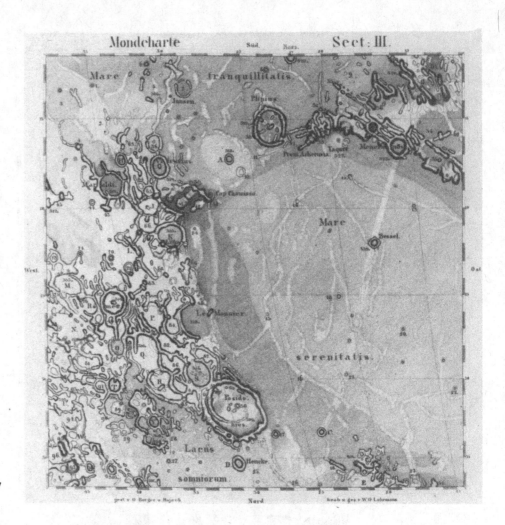

Figure 2.8
Section III of Wilhelm Lohrmann's map, drawn in 1824,
with additional nomenclature by
J. F. J. Schmidt (1878).

finest map ever produced until modern times (Figure 2.8). Unfortunately, the remaining twenty-one sections were not published until 1878 by J. F. J. Schmidt, by which time the three other major maps described below were available.

In 1834, only a decade after the publication of Lohrmann's four sections, W. Beer and J. H. Mädler completed their landmark map, followed in 1838 by *Der Mond,* a comprehensive treatment of the whole subject of lunar science. The book includes numerous measurements of mountain heights and crater depths, of crater diameters and coordinates, and detailed descriptions of the topography. The map, titled "Mappa Selenographica," echoes Lohrmann's in that it has the same diameter, the same 5-degree latitude and longitude intervals, and again uses hachures to portray topography; it was, however, produced in four sections (quadrants). Beer and Mädler followed Schröter's scheme of using Riccioli's nomenclature, augmenting it with almost all of Schröter's additions and some 140 names of their own choice. They also added Greek and Roman letters for designating peaks, mountain blocks, points with measured heights, rilles, ridges, and subsidiary craters associated with a large, named feature. Beer and Mädler's nomenclature advanced beyond Schröter's by employing a more convenient and logical scheme for the disposition of the letters.

The third influential lunar map produced by a German astronomer was that of Edmund Neison, published in 1876. It was never available as a single sheet or a separate set, but only as twenty-two overlapping sections that illustrated descriptive chapters in his large book *The Moon*. Both the map (62 cm in diameter) and descriptions rely heavily on Beer and Mädler's earlier work, but with many additions of subsidiary features and designations, plus some corrections. This map also uses the hachure technique but with less elegance than in its two predecessors. It also omits all albedo markings.

The end of the major nineteenth century mapping – and incidentally the end of the prephotographic era in lunar mapping – came with the publication of J. F. Julius Schmidt's great map in 1878, the same year that he published the remaining sections of Lohrmann's map. Using several larger telescopes than those available to his predecessors, he produced a highly detailed map (it depicts over thirty thousand craters!) in twenty-five sections at twice the scale of the Lohrmann and the Beer and Mädler maps. He used the hachure technique and indicated albedo differences by shadings with a sepia tone. The map is based on Lohrmann's and Beer and Mädler's position measurements, with eye estimates for positioning the intervening detail. Lines of latitude and longitude are not marked on the map, but 1-degree intercepts are provided in the margins where feasible. Features are given numbers or letters to prevent the masking of detail, the accompanying *Erläuterungsband* providing the names of designated features. Both Schmidt and Neison added new names, but they did not make full reference to each other, so some features have different names, and some names have two features!

THE PHOTOGRAPHIC ERA   The advent of practical photography in astronomy during the last decade of the nineteenth century had a profound effect on lunar cartography. First of all, it meant that positions of lunar features could be determined with much greater accuracy and speed – and with much less effort – than by making direct measurements through the telescope. Second, surface details could be mapped with much greater ease and accuracy. A third effect was that professional astronomers turned to the rich fields of stars, nebulae, galaxies, and so forth that were made available by long-exposure photography, relegating lunar cartography and studies almost entirely to amateur astronomers.

Position measurements were soon undertaken by J. Franz in Germany and S. A. Saunder in England, the former concentrating more on features situated near the edge of the Moon, whereas the latter dealt with the more central regions. Their final catalogs appeared in 1913 and 1911, respectively, and included nearly forty-five hundred measurements. Franz (1913) worked exclusively in selenographic latitudes and longitudes (a complex mathematical procedure), but Saunder used direction cosines, a system devised by H. H.

Turner (Saunder, 1911) that has several advantages. In this system, any point on the lunar surface has three orthogonal coordinates – xi, eta, and zeta – that correspond to $x, y,$ and z in cartesian coordinates, with the xi–eta plane passing through the Moon's poles and the mean east and west limbs. The radius is reckoned as unity. Accurate cartographic plotting with positions given in latitude and longitude was virtually impossible near the limb because of the rapidly changing scale in longitude, but it is very simple and accurate with xi and eta coordinates.

In 1910 W. Goodacre, an English amateur, was the first to draw a map using Saunder's early position measurements in rectangular coordinates. The format is similar to that of Lohrmann and Schmidt: twenty-five sections, south "up," with a diameter of 196 cm. Topographic details are indicated by means of form lines in which crests, bottoms of slopes, and other less obvious land-forms are represented by plain lines. This involves much less work and artistic ability than does hachuring, but it introduces ambiguities in interpretation. Albedo differences are not included; nor are lines of latitude or longitude. The map was intended chiefly as a guide for amateur observers. It was published in 1931 at a scale of 76 cm to the lunar diameter as an illustration for Good-acre's book *The Moon* (Figure 2.9).

An important map, but one with a different purpose, was published in 1935 (Blagg and Müller, 1935). This was the International Astronomical Union (IAU, founded in 1919) map, which was accompanied by a separate catalog of nomenclature and positions. As noted earlier, nomenclature had become very confused by the close of the nineteenth century, and an early task of the IAU was to establish standards. Mary Blagg had addressed this problem a decade earlier, and the 1935 publication is almost entirely her work. The map is heterogeneous, the inner partial quadrants having been nicely drawn by an artist (W. H. Wesley), the ten outer sections by Blagg herself at a different scale and with notably less clarity. The nomenclature was agreed upon by a small IAU committee, following Blagg's recommendations. The actual lettering, which includes many hundreds of subsidiary formations, was also done by Blagg and is not always as legible or unambiguous as one would desire. Thus, although the main intent of the map was to standardize and illustrate the nomenclature, it did not fully succeed in that task. Rectangular (i.e., xi–eta) coordinates, taken from the final catalogs of Saunder and Franz, the usual south-up orientation, and a lack of albedo indications characterize the Blagg–Wesley map.

The only other widely available map from the "Earth-based" era was the so-called 300-inch map by H. P. Wilkins, another British amateur astronomer. It was published in twenty-five sections at 100-inch scale during the 1950s but is more widely known as illustrations to the book *The Moon* by Wilkins and P. A. Moore (1961), in which it appears at a scale of 83 cm to the lunar diameter. The map follows the Goodacre format and mapping style but is even less artistic and is cluttered with detail, much of which is uninterpretable or spurious. In addition, it has maps and charts of the limb and libratory

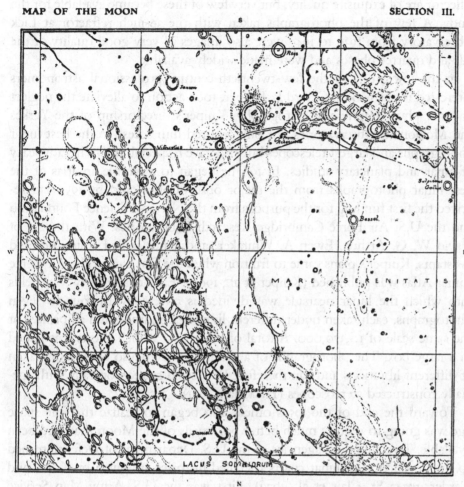

MAP OF THE MOON   SECTION III.

Drawn by WALTER GOODACRE, F.R.A.S., 1910.

Figure 2.9
Section III of W. Goodacre's map, drawn in 1910 (Goodacre, 1931).

regions, but unfortunately these are mostly not even self-consistent. The nomenclature does not rigidly follow that of the IAU, and it introduces over ninety new names that are "unapproved" (by the IAU). However, because nothing better was available, Soviet scientists in 1959 were obliged to use these maps and limb area charts to identify limb features photographed by the Luna 3 spacecraft for extrapolation to the previously unseen areas of the farside that also appeared in the Luna 3 images.

As an adjunct to mapmaking or simply for direct study, lunar photographs did not attain useful resolution and quality until the close of the nineteenth century. The only collection with adequate scope and resolution that was available prior to 1960 was Loewy and Puiseux's *Atlas Photographique de la Lune,* usually referred to as The Paris Atlas. It was published in twelve sections from about 1896 to 1909 and contains seventy-one large, grossly overenlarged sheets that cover virtually the entire nearside under various illumination angles. The average resolution is about that of a 3-inch telescope used visually. Many photographs taken during the 1919–27 era with the 100-inch Mt. Wilson reflector are of exquisite quality, but very few of these became available for the study. A few of the photographs taken with the 36-inch refractor at Lick Observatory from 1936 to 1947 were also images of very good quality of the full and quarter phases and were made widely available.

In the 1950s, one of the few twentieth-century professional astronomers interested in the Moon, Gerard P. Kuiper, took action to alleviate the neglect of lunar photography and cartography. Kuiper's directorship of the Yerkes and McDonald Observatories not only assured him access to the best lunar photographs and largest telescopes but also gave respectability to his advocacy of lunar and planetary studies. In 1955 he began to assemble an atlas of the best lunar photographs from the major observatories, and by 1957 had received the first funding for the purpose from the National Science Foundation and the U.S. Air Force Cambridge Research Laboratories. With the aid of David W. G. Arthur, Ewen A. Whitaker, and other skilled colleagues and assistants, Kuiper's plans came to fruition when the first part of the Air Force Lunar Atlas was published (Kuiper et al., 1960). Each of the forty-four fields into which the lunar nearside was divided is covered with four to seven photographs, each taken under different lighting condition and all printed at the same scale of 1:1,370,000. A total of 230 17- × 21-inch sheets was issued in a large box. The photographs of a given area were also commonly taken at different librations and can therefore be viewed stereoscopically with specially constructed stereoscopes (Hackman, 1961).

Toward the end of the 1950s others also began to realize that the space age was going to require much better portrayals of the Moon than had been available. Two mapping agencies of the U.S. Department of Defense turned to the production of lunar charts (detailed histories are given by Kopal and Carder, 1974; St. Clair et al., 1979). First was the U.S. Army Map Service (AMS), which in late 1958 began an effort to produce topographic maps of the nearside by stereophotogrammetric techniques. This effort was doomed to failure by inadequate data and inadequate stereoplotters, and, in fact, no such map has yet been made of the entire nearside. Nevertheless, between 1962 and 1965, AMS published twelve maps with a total of twenty-seven sheets at scales of 1:2,000,000 to 1:5,000,000, and even attempted a four-sheet topographic map at the very large scale of 1:250,000.

A much more successful and enduring program of lunar mapping was begun in the fall of 1959 by the U.S. Air Force Chart and Information Center (ACIC), St. Louis, under the direction of Robert W. Carder. The centuries-old difficulty of portraying lunar features realistically was finally overcome when Patricia M. Bridges successfully prepared the first ACIC lunar chart by the artistic technique of airbrushing (Figure 2.10). After some help from Kuiper's group at Yerkes, ACIC began prolifically to create the lunar charts that were the basic cartographic data for the exploration of the Moon. The Moon was divided into 144 1:1,000,000-scale map segments called Lunar Astronautical Charts (LAC), forty-four of which were on the central nearside, which would be the target of most spaceflights. In 1960 ACIC also published

Figure 2.10
*P. M. Bridges, preparing proto-
type airbrush map of Copernicus
area to demonstrate technique,
1959 (this map was prepared at
ACIC, St. Louis, in secret, to
prevent discovery by AMS, who
was expected to compete for exclusive rights to lunar mapping).*

the first edition of the much-used orthographic Lunar Reference Mosaic
(Lunar Earthside Mosaic, or LEM), which was later revised and published at
three scales.

The charts by AMS and ACIC incorporated a new "astronautical" convention of lunar directions. These had been inverted on many early lunar maps,
beginning with Gilbert's drawing of about 1600 (Figure 2.1), in which north
is "up" but the east and west directions from the terrestrial observer were
used. For example, the feature labeled Regio Magna Orientalis on the Gilbert
sketch roughly corresponds in position to Mare Imbrium, whereas Regio
Magna Occidentalis appears to be Mare Serenitatis and Mare Tranquillitatis.
Continuation of this convention resulted in the naming of Mare Orientale
("Eastern Sea") on the Moon's west limb. At its 1960 meeting, the IAU
decided to define lunar directions so that an astronaut would see the Sun rise
over the eastern horizon on the Moon as he would on Earth (IAU, 1960).
Furthermore, north would be "up" in subsequently published lunar illustra-
tions and maps.

In 1961 cartographer Bridges and observers William D. Cannell and James
Greenacre moved to Flagstaff, Arizona, where they could integrate studies of
photographs with visual observations made through the 24-inch refractor of
Lowell Observatory. The locations and general appearance of features were
obtained from photographs, the best of which could resolve objects 300 to
400 m across. Details as small as 200 m across were added during brief
moments of excellent "seeing" that could be captured only by the eye. The

group, led by Cannell, expanded considerably in the next few years. Between 1960 and 1967 they completed, almost entirely from telescopic data, the relief portrayals of the forty-four nearside LACs and also of twenty 1:500,000-scale Apollo Intermediate Charts (AIC) in the equatorial belt, which was of primary interest for early Apollo landings. The LACs are still the standard lunar charts, and only a few of them have been superseded by others using space age image data. Airbrushing has continued to be a prime technique for portraying lunar and planetary surfaces (see Chapter 3).

At the time this mapping was begun, the National Aeronautics and Space Administration (NASA) and ACIC adopted an illumination convention for lunar shaded relief maps. Illumination from the northwest had been an established convention for terrestrial relief shading, based on a theory that most people see relief images correctly, rather than inverted, when they are illuminated from the upper left. That is, mountains appear to be mountains rather than depressions. This convention was modified for lunar maps, because most lunar photographs are illuminated either from the east or the west; westerly illumination was therefore selected as the closest compromise between available data and tradition. Several landing site maps were later made with easterly illumination when it was determined that Apollo would land during the lunar morning, on the assumption that this would make them easier for the astronauts to use. Westerly illumination has been reestablished for all planetary mapping.

Other agencies cooperated with ACIC in the production of these vital lunar charts. Lowell and Pic du Midi Observatories were actively engaged in photography for the cartographic effort, and a half dozen other major observatories also intermittently contributed excellent lunar photographs that were incorporated. Relative elevations were computed by shadow measurements performed by a group directed by Zdenek Kopal at the University of Manchester, England. Elevations of many points were printed on the charts and were also used to derive contours, which, however, are reliable only in local areas. The geodetic (selenodetic) control and feature nomenclature for the charts were supplied by the Lunar and Planetary Laboratory (LPL), established by Kuiper at the University of Arizona in 1960. LPL was the first modern institution devoted primarily to lunar and planetary studies. In its role as a provider of basic data for lunar and planetary studies, LPL also prepared a four-part catalog and four-quadrant chart of measured, positioned, and named lunar features (Arthur and Whitaker, 1960; Arthur et al., 1963a, 1963b, 1965, 1966; Arthur and Agnieray, 1964; Arthur and Pellicori, 1965). These were accepted by the IAU as the official successors to the 1935 map and catalog of Blagg and Müller. LPL also continued work on the Air Force Lunar Atlas during the 1960s. In 1963, the first part of the atlas was supplemented by bound versions in which controlled graticules were added to rectified photographs (Whitaker, et al., 1963; Figure 2.11). It has remained unique and useful despite the availability of spacecraft photographs. The rectifications were made by the simple but effective technique of rephotographing projec-

Figure 2.11
Rectified photomap of Copernicus area (Whitaker et al., 1963).

tions of telescopic photographs on a large, blank globe. In 1965 LPL began systematic photography that was tailored to lunar purposes, using a new 1.55-m telescope located in the Catalina Mountains north of Tucson. In 1967 the best of these photographs, along with a full-Moon series taken with another 1.55-m telescope at the Naval Observatory at Flagstaff, were published together (in a box) as the third and fourth supplements to the Air Force Atlas in 1967 (Kuiper et al., 1967). To obtain the best possible reproduction, the photographs were reproduced photographically on glossy photographic paper, a method of publication that necessarily limited this still-valuable atlas to a small edition. The decade-long work on the Air Force Atlas ended as the era of systematic photographic spaceflight began, but the nascent space age had obtained its first widely available and usable collection of lunar photographs. This atlas also contains index maps of Lunar Orbiter photographs.

Another, almost totally new type of lunar mapping, geologic, began in the milestone year of 1959. A number of geologists, including the American Spurr, the German Kurd von Bülow, and the Russian A. V. Khabokov had attempted geologic maps of the Moon, but these were based mostly on analyses of linear features that their authors assumed to be of internal origin. Thus, they were interpretive structural or tectonic maps. A true geologic map portrays the distribution and age relations (the stratigraphy) of three-dimensional rock bodies known as geologic units (see Chapter 7). To our knowledge, the first map that portrayed lunar stratigraphic units was one of the four sheets of the "Engineer Special Study" prepared in 1959 and 1960 by Robert J. Hackman and Arnold C. Mason of the U.S. Geological Survey (USGS) (Hackman and Mason, 1961; Hackman, 1961; Mason and Hackman, 1962).

Though conceptually modern, this pioneering map showed only three stratigraphic units: premare, mare, and postmare. A further step was taken later in 1959 by Eugene M. Shoemaker. He happened to see the ACIC prototype LAC and immediately used it as a base for a geologic map of the Copernicus region, which he had been studying for other purposes (Shoemaker, 1962a; Figure 2.12). Soon afterward, a landmark paper by Shoemaker and Hackman (1962) presented a nearly complete lunar stratigraphic scheme and explained the principles of lunar stratigraphic analysis and geologic mapping.

Geologic mapping became the focus of a program of astrogeologic studies that was initiated in the USGS in August 1960 and acquired formal status as a Branch in September 1961 (Shoemaker, 1981). Shoemaker realized that geologic maps would be needed for selection of scientifically productive and safe exploration sites and for the extrapolation of data from these necessarily small sites to the rest of the Moon. The forty-four planned ACIC 1:1,000,000-scale LACs were the base maps for the first phase of this mapping (see Appendix I). The first four LACs to be geologically mapped constituted the "Lansberg" region, an airbrush map that ACIC published in a limited and now-rare edition at the scale of 1:2,000,000. At the time, the Lansberg region was considered for the target area of the early Ranger and Surveyor spacecraft, and it was furthermore thought that the Apollos would land at spots previously tested by these unmanned missions (Shoemaker, 1962b). Thus, the four geologic "quadrangle" maps were to be a test of mapping for exploration purposes. They included Shoemaker's prototype map of the Copernicus region (LAC 58, published formally only as black-and-white text; Figure 2.12), and later maps of LACs 75 and 76 (Marshall, 1963; Eggleton, 1965). The mapping program became more intensive and elaborate at the approach of the end-of-the decade deadline set in May 1961 by President Kennedy for the first manned lunar landing. By mid-1966, twenty-eight quadrangles had been prepared in preliminary form, although only eight had been formally published (see Wilhelms, 1970; Wilhelms and McCauley, 1971). The geology of the equatorial belt between 32° N and S latitudes, embracing the twenty-eight LACs, had also been summarized at the 1:5,000,000 scale (Wilhelms et al., 1965). In 1967 photographs obtained by the Lunar Orbiters (especially Orbiter 4) obviated most of the need for further telescopically based work, and subsequent lunar geologic maps have been prepared largely on the basis of spacecraft photographs.

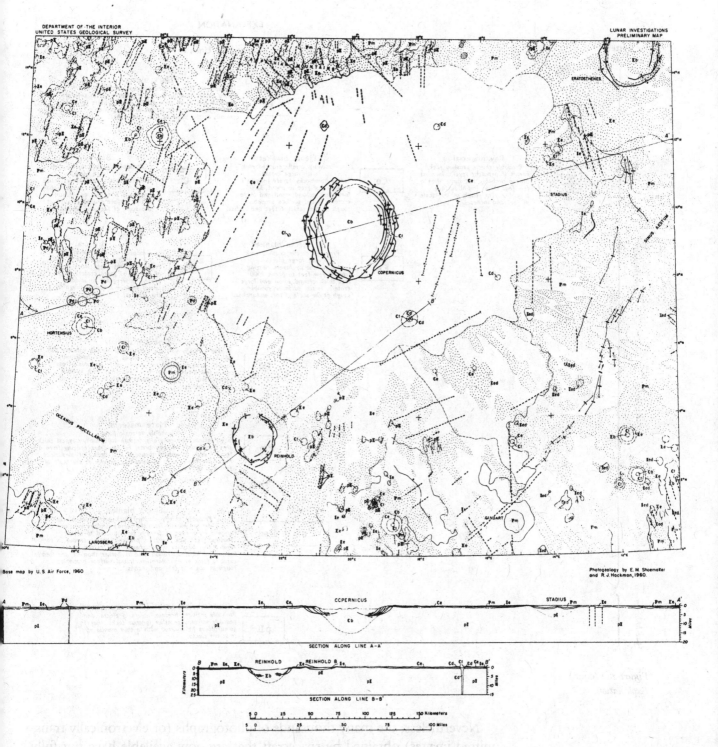

Figure 2.12
*Prototype geologic map of Copernicus area (Shoemaker, 1962a). (*Explanation shown on following pages.)

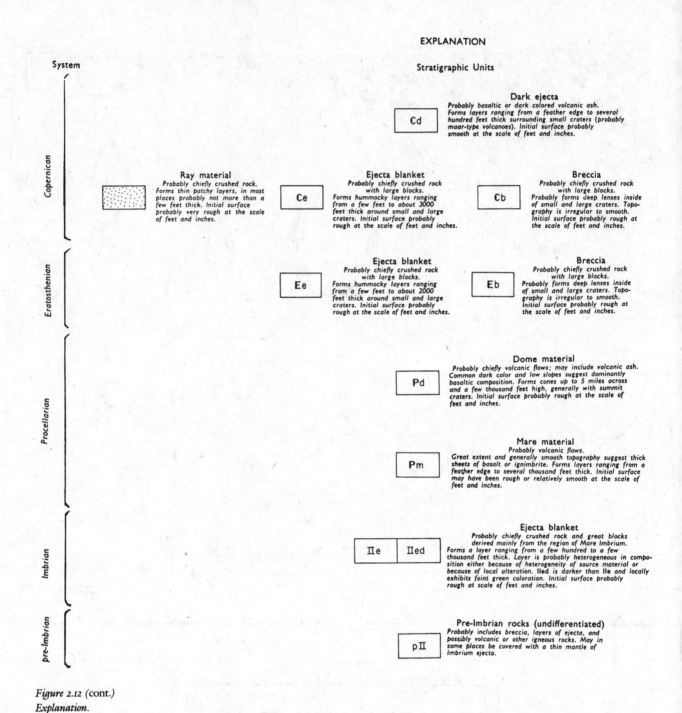

EXPLANATION

System

Stratigraphic Units

Copernican

Ray material
Probably chiefly crushed rock. Forms thin patchy layers, in most places probably not more than a few feet thick. Initial surface probably very rough at the scale of feet and inches.

Ce — Ejecta blanket
Probably chiefly crushed rock with large blocks. Forms hummocky layers ranging from a few feet to about 3000 feet thick around small and large craters. Initial surface probably rough at the scale of feet and inches.

Cd — Dark ejecta
Probably basaltic or dark colored volcanic ash. Forms layers ranging from a feather edge to several hundred feet thick surrounding small craters (probably maar-type volcanoes). Initial surface probably smooth at the scale of feet and inches.

Cb — Breccia
Probably chiefly crushed rock with large blocks. Probably forms deep lenses inside of small and large craters. Topography is irregular to smooth. Initial surface probably rough at the scale of feet and inches.

Eratosthenian

Ee — Ejecta blanket
Probably chiefly crushed rock with large blocks. Forms hummocky layers ranging from a few feet to about 2000 feet thick around small and large craters. Initial surface probably rough at the scale of feet and inches.

Eb — Breccia
Probably chiefly crushed rock with large blocks. Probably forms deep lenses inside of small and large craters. Topography is irregular to smooth. Initial surface probably rough at the scale of feet and inches.

Procellarian

Pd — Dome material
Probably chiefly volcanic flows; may include volcanic ash. Common dark color and low slopes suggest dominantly basaltic composition. Forms cones up to 5 miles across and a few thousand feet high, generally with summit craters. Initial surface probably rough at the scale of feet and inches.

Pm — Mare material
Probably volcanic flows. Great extent and generally smooth topography suggest thick sheets of basalt or ignimbrite. Forms layers ranging from a feather edge to several thousand feet thick. Initial surface may have been rough or relatively smooth at the scale of feet and inches.

Imbrian

IIe / IIed — Ejecta blanket
Probably chiefly crushed rock and great blocks derived mainly from the region of Mare Imbrium. Forms a layer ranging from a few hundred to a few thousand feet thick. Layer is probably heterogeneous in composition either because of heterogeneity of source material or because of local alteration. IIed is darker than IIe and locally exhibits faint green coloration. Initial surface probably rough at scale of feet and inches.

pre-Imbrian

pII — Pre-Imbrian rocks (undifferentiated)
Probably includes breccia, layers of ejecta, and possibly volcanic or other igneous rocks. May in some places be covered with a thin mantle of Imbrium ejecta.

Figure 2.12 (cont.)
Explanation.

Nevertheless, the many thousands of photographs (or electronically transmitted images) obtained by spacecraft that are now available have not fully replaced those obtained by telescopes. Telescopic views usually cover larger areas and therefore provide the "big picture" that is essential for geologic and other scientific analyses. The low Sun illuminations provided by telescopic photographs that include the terminator were only rarely obtained by spacecraft, so that subtle features such as low mare domes and ridges are often more easily seen telescopically. Furthermore, features like crater rays and subtle

Surface Characteristics

Generally smooth at the scale of miles.
Topography controlled by relief on
contact with underlying material.
Probably smooth at the scale of feet and
inches. Among the youngest material on
the lunar surface; initial surface
characteristics probably largely un-
modified.
Low reflectivity.

Talus
*Probably partially sorted accumu-
lation of fragments ranging in
size from dust to large blocks.
Generally forms sheets mantling
smooth slopes of about 30°. Initial
surface probably rough at the
scale of feet and inches.*

Ct

Topography at the scale of miles varies
from pitted to hummocky to smooth.
Probably rough to very rough at the
scale of feet and inches; initial surface
characteristics probably fresh to partially
modified.
High to very high reflectivity

Rill and chain crater material
*Probably includes breccia, fault blocks, and volcanic
rocks or serpentine. Age not definitely established
but probably chiefly Eratosthenian.*

Topography at the scale of miles varies
from pitted to hummocky to smooth.
Probably smooth to slightly rough at the
scale of feet and inches; initial small
scale relief probably much reduced by
small meteorite bombardment, insolation
and mass movement.
Low to moderate reflectivity.

Contact
Dashed where approximately located

Indefinite contact

Fault
Dashed where approximately located
U, upthrown side; D, downthrown side

Volcano-shaped topographic forms.
Probably smooth at the scale of feet
and inches; initial small scale relief
probably largely reduced by small
meteorite bombardment, insolation and
mass movement.
Low to moderate reflectivity.

Concealed fault or fracture
Queried where probable

Anticline
*showing trace of axial plane and bearing
and plunge of axis*

Smooth plain broken by small meteorite
and secondary impact craters and
rounded ridges of several hundred
feet relief.
Probably smooth at the scale of feet
and inches; initial small scale relief
probably largely reduced by small meteorite.
Low reflectivity.

Syncline
showing trace of axial plane

Flow front or monocline
showing direction of slope of surface scarp

Hilly to locally smooth at the scale of
miles. Topography characterized by
numerous hills and depressions one to two
miles across; locally controlled by relief
on contact with pre-Imbrian rocks.
Probably smooth at the scale of feet and
inches; initial small scale relief probably
largely reduced by small meteorite bombardment, insolation, and mass movement.
Low to moderate reflectivity.

Not well exposed.

albedo contrasts, which show up best under the opposite kind of illumination
(very high, that is, near full Moon), are also often best viewed on telescopic
photographs.

2.1.2 *Planetary cartography and photography (1840–1965)*

Because of their small angular size as viewed from Earth, the planets present
far greater problems for the cartographer than does the Moon. Indeed, just

the visual depiction or photography of the small disks by ground-based telescopes is no trivial challenge even today. Markings on Mars and Jupiter were recorded in the first half of the seventeenth century, but the other planets present only cloudy, essentially featureless atmospheres to us, and thus were not amenable to mapping. In fact, the only planets that show solid surfaces are Mercury and Mars (Pluto is starlike), and even Mars has occasional clouds and dust storms that obscure detail and confuse the observer. Furthermore, not all markings on Mars are permanent, adding further problems for the cartographer.

MERCURY This planet is notoriously difficult to observe effectively because of its small angular size and perpetual proximity to the Sun as seen from Earth. The first meaningful observations of surface markings were made in the 1880s by G. V. Schiaparelli. From his observations he concluded that the planet rotated in the same period as its revolution around the Sun – about eighty-eight days – and that the axis was closely, if not exactly, perpendicular to the orbital plane. He was thus able to draw a map, the zero of longitude being the subsolar point at aphelion and perihelion. With these assumptions, only about 226 degrees of planetary longitude ever saw sunlight, the extra 23 degrees on each side of the mean limbs arising from librations due to Mercury's eccentric orbit. Schiaparelli's map (1881), which is circular, only runs from $-90°$ to $+90°$ longitude and employs an unusual projection in which the ten divisions are equally spaced along both equator and prime meridian (Antoniadi, 1933; Figure 2.13).

The following half-century saw the production of numerous maps, that of E. M. Antoniadi gaining the most prominence (Figure 2.14). He added a limited nomenclature for descriptive purposes. The synchronous rotation/revolution assumption was rudely shattered in 1965 when radar observations gave a rotation period of about 58.6 days, two-thirds of the revolution period. This threw extreme doubt on the validity of the earlier maps, but it was soon realized that the fortuitous near-commensurability between favorable planetary apparitions and the fifty-nine-day rotation period could, given the observational difficulties, effectively mask the fact that the rotation period was fifty-nine days and not eighty-eight. However, the fifty-nine-day period meant that there are two perihelial subsolar points, separated by 180 degrees, and two aphelial ones situated 90 degrees from the other two!

The first map based on the new rotation period was compiled by D. P. Cruikshank and C. R. Chapman in 1967 from a selection of the older drawings. They used a cylindrical projection with north "up" but defined a new zero of longitude. More recent compilations, using photographs made at the Pic du Midi and New Mexico State University observatories, and visual observations from the former, have been combined into a single map on a Mercator projection. The zero of longitude is the same perihelion subsolar point as used by Schiaparelli and Antoniadi.

Radar provides an effective technique for measuring slopes from the Earth.

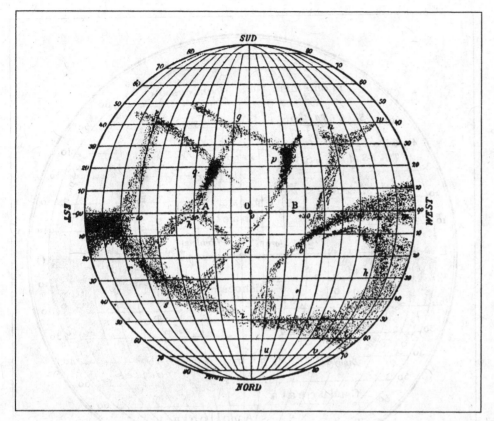

Figure 2.13
Map of Mercury by Schiaparelli,
circa 1889 (from Antoniadi,
1934).

In the early 1970s, G. H. Pettengill and Richard Jurgens derived altimetry profiles from Earth-based radar images of the equatorial zone of Mercury. Although these data provide valuable cartographic information, coverage was not complete and an accurate datum was not established. They have therefore never been integrated with shaded relief maps compiled from spacecraft data.

MARS Unlike Mercury, Mars is relatively easy to observe in some detail. Markings on it were first noticed by Francisco Fontana in 1636. In 1659 Huygens (1925) was first to sketch recognizable detail on Mars, showing the dark marking now known as Syrtis Major (Figure 2.15). Improvements in telescopes were accompanied by increases in both the quality and quantity of Mars drawings, but not until 1840 was the first complete map drawn and published – by Beer and Mädler of lunar-map fame. Drawing such a map entails determining the locations of the poles, hence the axis and equator, in relation to the surface features and also choosing a prime meridian. Mädler did this, the zero of longitude of his choice remaining unchanged (except for refinements) to this day. He marked longitudes positive toward areographic east, but this was reversed by Schiaparelli in 1877 so that central longitudes, as seen from Earth, increased with time. The map consists of northern and southern hemispheres presented in stereographic projection; only albedo markings are mapped, and no nomenclature was supplied.

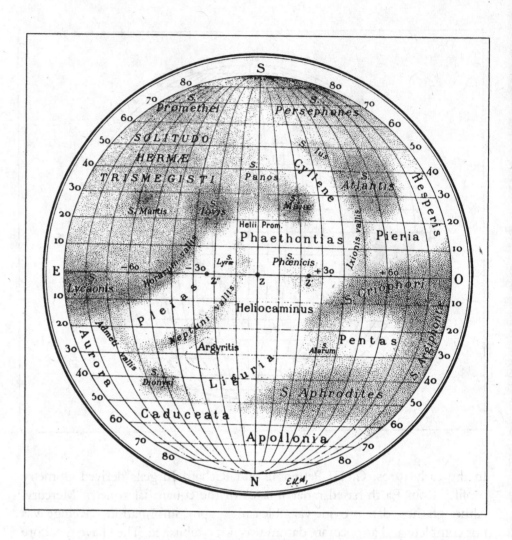

Figure 2.14
Map of Mercury by Antoniadi,
1933 (from Antoniadi, 1934).

The next maps to be published of Mars were those of R. A. Proctor in 1868 and 1871. They are based on observations by several experienced observers and contain more detail than the Beer and Mädler map, but they are crudely drawn and lack aesthetic appeal. The first of these maps uses Mercator and stereographic (Figure 2.16) projections, whereas later maps by Proctor use the so-called equidistant projection. Proctor was the first to employ a nomenclature scheme for the planet, using the names of current and earlier astronomers in conjunction with terrestrial analogs, such as ocean, sea, inlet, continent, and land. Ill-advisedly, he sometimes used the name of the same observer for several different landform terms, such as Dawes Sea, Dawes Island, Dawes Continent, which contributed to the demise of the system (Proctor, 1892).

G. V. Schiaparelli was the first, as with Mercury, to produce a definitive map of Mars. He observed the planet during the very favorable apparitions of 1877 and 1879, making drawings and position measurements from which he produced a map on the Mercator projection (Figure 2.17). He rejected

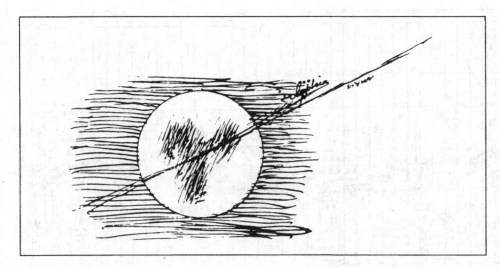

Figure 2.15
Drawing of Mars by Huygens,
1658 (Huygens, 1925).

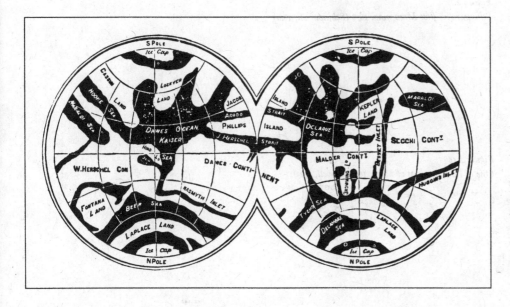

Figure 2.16
Proctor's first map of Mars on a
stereographic projection. This
map was probably drawn in
1868, and was published as a
map sheet in 1871. This illustra-
tion was printed in Proctor
(1892).

the Proctor nomenclature and a similar one by Flammarion, instead using historical and mythical names from classical literature and geography. This scheme has stood the test of time and formed the basis of the 1958 IAU Mars map.

It was soon realized that the surface markings were not constant; not only would apparent obscurations of one type or another occasionally mask various areas, but some details would change appearance or vanish or appear. This was a problem for cartographers, who were unable to draw a definitive map; each apparition resulted in a slightly different map! Another problem was that of the "canals," evanescent narrow lines that seemed to form a network on the Martian surface. Not all observers saw these features, and those who did seldom agreed on their positions. Thus, many maps appeared in the half-

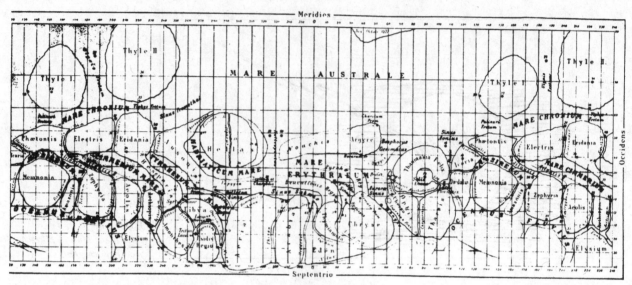

Figure 2.17
Schiaparelli's first Mars map, 1878 (Schiaparelli, 1881).

Figure 2.18
One of Lowell's maps of Mars, 1901 (Lowell, 1923).

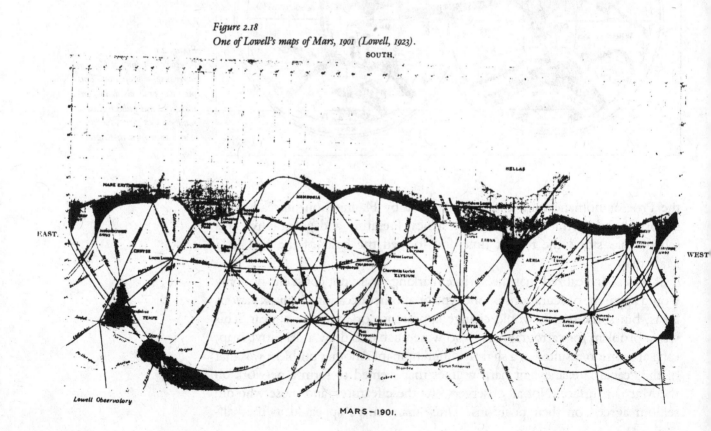

century following Schiaparelli's first maps, the most influential being those of E. M. Antoniadi, a careful observer and excellent artist. The maps by P. Lowell (1923) (Figure 2.18), with their complicated network of "canals" intersecting "oases," caught the public's imagination, with the implication of intelligent beings at work, but they were judged to be scientifically inaccurate. The Mercator projection became standard, with south "up" and longitudes increasing to areographic west; i.e., longitudes of the subsolar and sub-Earth points increased with time. The zero of longitude chosen by Beer and Mädler and followed by Schiaparelli also became standard and has been moved only about 3 degrees on modern maps based on images from spacecraft.

As noted for the Moon, photographs of planets rarely show as much detail as can be seen visually through the same telescope. However, they are objective records and, with appropriate precautions, can yield good positional data. They are also invaluable as records of temporal events and permanent changes.

The series of photographs taken by E. C. Slipher at Lowell Observatory, beginning in 1905, thus forms an invaluable resource, augmented by photographs from other observatories. The excellent maps drawn by G. de Mottoni in the 1950s were based on photographs taken at the Pic du Midi Observatory; these were used as the basis for the 1958 IAU general map of Mars.

VENUS Radar telescopes at Goldstone, California (Goldstein et al., 1978), and Arecibo, Puerto Rico (Campbell and Burns, 1980), have been used to make radar reflectivity images of Venus. Although these images are theoretically capable of resolving features as small as 20 to 40 km, the viewing geometry is such that topography is not resolved directly. The bright areas on the images are evidence of rough surfaces, and the darker areas indicate smoother surfaces. This information, when correlated with image and altimetry data returned by spacecraft, is capable of yielding significant geological and tectonic information.

The radar images have been reconstituted in the Mercator projection to provide illustrations for reports and scientific articles (Figure 2.19), but maps made solely from Earth-based radar imaging have not been published.

2.2. LUNAR AND PLANETARY MAPPING FROM SPACECRAFT DATA

2.2.1. The Moon

Despite the good quality of telescopic photographs of the Moon, spacecraft data were required for closeup views of the surface and regional views of the farside. The U.S. lunar program succeeded in obtaining some photographs despite a primary interest in closeup data directly relevant to landing safety. Neither the United States nor the Soviets ever attempted to acquire uniform and comprehensive photographic coverage of the entire Moon.

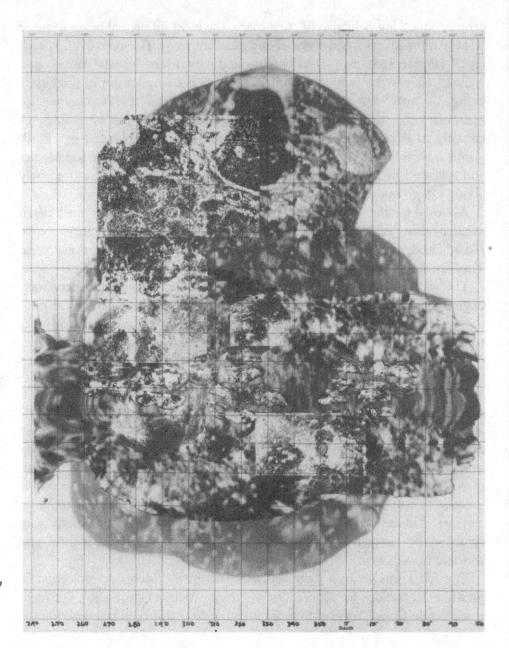

Figure 2.19
Radar map of Venus Earth-facing hemisphere, taken from the 1000-foot-diameter radar telescope at Arecibo, Puerto Rico, and by NASA's three 64-m Deep Space Network Antennas at Goldstone, California (NASA photo).

LUNAS 1 TO 8 (1959–65) The Soviet lunar program has employed two series of spacecraft, Luna and Zond, each with widely varying design and objectives. The first and last lunar missions were called Luna. In January 1959 Luna 1 became the first mission to attempt to reach the Moon but missed by 6,000 km. Luna 2 impacted the rim of the crater Autolycus in September 1959 but returned no photographs.

The first spacecraft to return lunar photographs was Luna 3, which in October 1959 provided the first view of the previously unknown lunar farside (Figure 2.20A and B). Although the high-illumination ("full-Moon") view

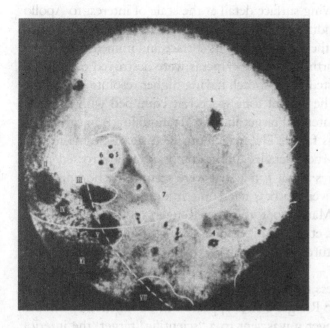

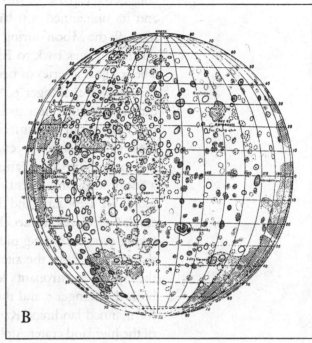

A

B

Figure 2.20

A. *First spacecraft image of the lunar farside, taken by Soviet Luna 3 in 1959.* Farside features: *1. Sea of Moscow; 2. Gulf of Cosmonauts; 3. Continuation of Mare Australe; 4. Tsiolkovsky Crater; 5. Lomonosov Crater; 6. Joliot-Curie Crater; 7. Sovietsky Mountain Range; 8. Sea of Dreams.* Nearside features: *I. Mare Humboldt; II. Mare Crisium; III. Mare Marginis; IV. Mare Undarum; V. Mare Smythii; VI. Mare Fecunditatis; VII. Mare Australe. B. Map of the lunar farside, made by the Soviets from the Luna 3 image above.*

was of poor quality, it nevertheless revealed the important fact that the farside has fewer maria than the nearside (Whitaker, 1963). The Soviets published an atlas of maps based on the new data (Barabashov et al., 1961). Later missions would rectify some early misconceptions and clarify the nature of some of the newly observed markings, showing, for example, that the "Soviet Mountains" were actually coalescing rays of two young craters.

Lunas 4 through 8 were flown between 1963 and 1965 in soft-landing experiments (Woods, 1981). Luna 4 (April 1963) missed the Moon by 8,500 km, and passed into a 90,000- × 700,000-km Earth orbit. It returned no cartographic data. Luna 5 (May 1965) was the first soft-landing attempt, but it crashed in Mare Nubium without returning data. Luna 6 (June 1965) missed the Moon by 160,000 km. Lunas 7 and 8 (October and December 1965) crashed in Oceanus Procellarum.

RANGER (1961–65)   As an answer to the Luna 3 flyby and other early Soviet successes, the United States began its assault on the Moon; Project Ranger was initiated in December 1959 (Hall, 1977). The first five Rangers (August 1961 through October 1962) failed to return data. The scientific payload for the remaining Rangers (6 to 9, "block 3") consisted of the one instrument considered capable of showing surface detail at the scale of interest to Apollo and its unmanned soft-landing precursor, Surveyor. Rangers were to photograph the Moon during their approach to its surface and immediately transmit the images back to Earth before the cameras were destroyed on impact. The result is a series of nested photos, each having higher resolution than the previous one. Ranger 6, the first of four spacecraft equipped with the new cameras, crashed near its intended target in Mare Tranquillitatis in early February 1964, but its cameras failed. The first unqualified success came in July 1964 when Ranger 7 returned detailed photographs of a ray-crossed mare area before impact. Features about 1 m across were seen in the best pictures, a 200-fold improvement over the best telescopic resolution (Heacock et al., 1965). In February and March 1965 the last Rangers, 8 and 9, achieved similar successes (Heacock et al., 1966). Ranger 8 cruised over the highlands, taking ever-improving pictures, until it impacted on Mare Tranquillitatis, less than 70 km from the site – Tranquility Base – that was to be occupied by the Apollo 11 astronauts less than four and a half years later. As the maria at both the Ranger 7 and the Ranger 8 sites appeared to be sufficiently smooth for manned landings, Ranger 9 was sent to a "scientific" target, the interior of the highland crater Alphonsus, and provided the highest-resolution pictures of the lunar surface that had yet been obtained.

In 1964 and 1966 ACIC prepared a total of seventeen charts, called Ranger Lunar Charts (RLCs), of the three potential landing sites, at scales ranging from 1:1,000 to 1:1,000,000. The USGS used Ranger data to construct the first geologic maps at scales larger than the 1:1,000,000 LAC-quadrangle mapping; two geologic maps were constructed for each of the three sites. After the initial burst of interest in this work, however, it was set aside because Lunar Orbiter photographs proved to be infinitely better. These first maps based on Ranger data were not published until between 1969 and 1971 – well into the Apollo era.

The Rangers succeeded in their primary objective, however, by showing that lunar slopes are typically gentle and that selected surfaces were smooth enough for successful landings. Also, they contributed early in the lunar program to the concept that the type of geologic unit one can study is related to the scale of the data: Lunar Orbiter photographs and maps at regional scales are required to show the basic bedrock units, and closeup (Ranger) photographs and large-scale maps are required to show the regolith and features that contribute to its formation (Trask, 1972).

ZONDS 1 TO 3 AND LUNAS 9 TO 14 (1965–66) Between the flights of the last Ranger in March 1965 and the first Surveyor in June 1966 the Soviets, after an almost 6-year lull in successful lunar missions, added more firsts to the two Lunas in 1959 that had initiated lunar exploration. On July 20, 1965, Zond 3 completed low-resolution coverage of the lunar farside by transmitting the first views of the west limb, including the Orientale basin (Lipsky, 1965, 1969). Thus, the entire Moon was now covered by some kind of photograph.

(Zonds 1 and 2 were Venus and Mars missions, respectively, but returned no data.)

In February 1966 the Soviets achieved another first when Luna 9 made the first successful "soft" (nondestructive) landing on the Moon (Lipsky, 1966; Shoemaker et al., 1966). For three days the spacecraft transmitted panoramic views of the lunar surface and showed that the surface material is strong enough to support landings by spacecraft.

In April 1966, the USSR also became the first nation to fly a spacecraft, Luna 10, in orbit around the Moon. A similar but somewhat shorter mission was flown by Luna 11 in August 1966, overlapping in time with the first U.S. Lunar Orbiter. The first Soviet photographic orbital mission was flown by Luna 12 in October 1966. It employed television transmission of the images, as did the U.S. Lunar Orbiters.

Another successful soft lander, Luna 13, followed in December 1966 and closed the first phase of the Soviet soft-landing program. These missions apparently were confined to the western near-equatorial region of the Moon; Lunas 5, 7, and 8 crashed between 3° and 9° N latitudes and between 23° and 63° W longitudes, near the successful Luna 9 and 13 landings. Luna 14 (April 1968) was another orbital mission that apparently returned only engineering data. Thus, soft-landing and orbital missions overlapped in the Soviet program as they did in the U.S. program.

SURVEYOR (1966–68)   When the Surveyor Project was originally conceived by NASA, it ambitiously called for twenty spacecraft, the first seven of which were engineering models and the remainder scientific models. Each spacecraft was to carry about 160 kg of instruments, including two cameras for taking surface panoramas and one camera for obtaining pictures of the landing site during spacecraft descent. As launch-vehicle delays and other factors compressed the time that would intervene between the Surveyor and Apollo missions, the scientific spacecraft were deleted and the engineering models were modified for scientific purposes. Much of the purely engineering instrumentation was dropped, and the payload was cut to 30 kg, including only one survey camera.

Surveyors were launched between May 1966 and January 1968. Five Surveyors landed successfully and transmitted nearly ninety thousand pictures. Surveyors 1, 3, 5, and 6 landed successfully on mare sites (Shoemaker et al., 1969). Surveyors 2 and 4, targeted for Sinus Medii (long regarded as a likely Apollo landing site) malfunctioned and transmitted no data from the surface (Surveyor 6 was finally successful in landing at that site). After the four successes in potential Apollo landing zones on the maria, the north rim of the crater Tycho in the southern highlands was selected as a scientifically interesting site for Surveyor 7.

Large-scale (1:1 to 1:20) maps of each of the landing sites were made from the Surveyor pictures (Shoemaker et al., 1969), and a comprehensive atlas was prepared for the landing site of Surveyor 5 (Batson et al., 1974).

LUNAR ORBITERS I TO 5 (1966–67) The U.S. Lunar Orbiter program (Mutch, 1970) produced a more comprehensive database for lunar cartography than did all other means combined. Five Lunar Orbiter flights photographed almost 99 percent of the lunar surface. The Orbiter coverage of the nearside equals the accumulated telescopic data bank in areal coverage and generally exceeds it in resolution. Lunar Orbiter provides all the existing useful coverage of the farside except disconnected strips and patches provided by Zond and Apollo. Orbiter photographs, however, are inferior to the best Apollo orbital photographs in tonal range and suitability for stereophotogrammetry.

Each flight was designed to make 211 exposures. Each exposure produced a high-resolution H frame nested inside a medium-resolution M frame. After initial verification sequences, they were exposed in groups of overlapping M frames. Special sequences produced overlapping H frames. The five flights returned about 1,650 useful frames out of a maximum of 2,110.

The five spacecraft flew three different types of one-month missions. Orbiters 1, 2, and 3 flew at low altitudes and low orbital inclinations to the equator. The main objective was to search for potential Apollo landing sites in a nearside belt extending 5 degrees north and south of the equator and 45 degrees east and west of the central meridian. These were the "prime" (P) sites. About one-fourth of the frames were devoted to "film-set" sequences to keep the film system "limber" between the prime-site photography. These "supplementary," or "secondary," targets (S sites) included the farside and, starting with Orbiter 2, targets of scientific interest on the nearside. The low-altitude frames of Orbiter 1 were hopelessly smeared by failure of an image motion compensation device, but Orbiters 2 and 3 returned the high-resolution data on the prime sites that Apollo required.

This success released the last two spacecraft for different types of missions. Orbiters 4 and 5 orbited at inclinations of 85 degrees to the equator and could view most of the Moon. The high-resolution camera of Orbiter 4 photographed most of the nearside at resolutions of about 75 to 125 m, whereas M frames show the whole disk visible from the spacecraft and afford regional views from many new perspectives. Many of the frames taken early in the mission are hopelessly fogged because of a malfunction of the lens cap; otherwise, this mission was a resounding success and has proved indispensable for lunar cartography and geologic mapping.

Orbiter 5 operated at altitudes between 100 and 195 km, lower than Orbiter 4 but twice as high as the first three Orbiters. Twenty percent of Orbiter 5's exposures were taken to supplement coverage by Orbiters 1, 2, and 3 of potential Apollo landing sites. The remaining 80 percent were taken to fill in the farside coverage or to show scientifically interesting features. This last Lunar Orbiter was the most nearly malfunction-free of the five.

An intense cartographic effort was immediately applied to sites of interest to Apollo. Splitting the effort, AMS and ACIC generated both photomaps and airbrush maps of seventeen partly overlapping Lunar Orbiter photographic sites that covered eleven potential Apollo landing sites. Scales were 1:25,000 and 1:100,000. The impressive total of forty-six charts, consisting of a total of seventy-eight sheets, was produced in a little more than a year (Kopal and Carder, 1974). Concurrently, the USGS prepared geologic maps in direct support of potential manned landings at seven of the sites. Two of these were visited, by Apollos 11 and 12. Between 1969 and 1972 AMS and ACIC prepared photomaps and a few airbrush maps of twelve "scientific" sites photographed by Orbiters 3 and 5: Scales were 1:25,000 and (mostly) 1:250,000, and fifty-one sheets were printed though not widely distributed. The USGS prepared premission geologic maps (two sheets each) of three of these regions (Apollo 14, 15, and 17 sites) and also of the Apollo 16 landing site, all of which were based on Orbiter 4 and Apollo-orbital photography.

Of greater lasting interest and wider usage than the Orbiter 1, 2, 3, and 5 site maps are the small-scale charts made of the entire Moon by the most experienced ACIC cartographers. They constructed and revised these charts as each new mission sent back new images. By the middle of 1971 they had produced four charts at 1:2,750,000, nine at 1:5,000,000 (four regions, two editions of some), two at 1:10,000,000, and a globe.

A parallel effort by the National Geographic Society had a far wider distribution than these. Their two-hemisphere map on a Lambert Azimuthal Equal-Area projection proved very useful for both scientific and pictorial purposes, as has a revision prepared by Jay Inge of the USGS in cooperation with the cartographic staff of National Geographic. The Soviet Union's cartographers also have prepared small-scale charts and globes based on their own and the American data (see Kopal and Carder, 1974).

Most geologists prefer studies of larger regions than the Apollo sites. The Lunar Orbiters supplied the photographs necessary for these regional studies, as well as for the cartographic products. They provided most of the farside views of large areas and photographed the nearside at scales intermediate between the regional, often blurred telescopic views and the extreme closeup detail of Surveyor. Of the forty-four 1:1,000,000-scale geologic maps of the nearside, twenty-eight were prepared or revised based on Orbiter 4 data. The information from the forty-four maps was then compiled and revised for the production of the first large-area synoptic lunar geologic map to be published, the 1:5,000,000-scale geologic map of the nearside of the Moon (Wilhelms and McCauley, 1971), an elaboration and expansion of the earlier compilation of the near-equatorial belt (Wilhelms et al., 1965). Later, the 1:5,000,000 mapping was extended to the entire Moon. The north, south, east, west, and central far "sides" were analyzed in light of the Apollo results, but the mapping was still based mostly on Orbiter photographs (Scott et al., 1977; Wilhelms & El-Baz, 1977; Lucchitta, 1978; Stuart-Alexander, 1978; Wilhelms et al., 1979).

ZONDS 4 TO 8 AND LUNAS 15 TO 24 (1968–76) Zond 4 (March 1968) was a test mission, and Zond 5 (September 1968) was the first spacecraft to

43

circumnavigate the Moon; it carried insects, tortoises, and seeds and took photos of the Earth but evidently not of the Moon. It was recovered from the Indian Ocean, in the first sea recovery by the Soviet Union. The last three Zonds flew between 1968 and 1970, after the end of the American Lunar Orbiter program and after the Luna 10, 11, and 12 orbital missions. Zonds 6, 7, and 8 returned photographic film to Earth, the first unmanned missions to accomplish this feat. They looped once from the northern nearside to the southern farside before returning to Earth for a braked landing. These Zonds obtained good stereoscopic coverage of strips of the farside, regional coverage of parts of the west limb barely covered by the U.S. Lunar Orbiters, the only color lunar photographs taken by unmanned spacecraft, and photographs useful for constructing topographic profiles.

The next Soviet mission was the mysterious Luna 15, which flew to the Moon during the Apollo 11 mission. It apparently initiated a new and important type of soft-landing Luna mission, designed to return soil and rock samples to Earth. Luna 15, however, returned no scientific data after landing or crashing at 12° N latitude, 60° E longitude. Three successful missions of this type, Lunas 16, 20, and 24, followed later (September 1970, February 1972, and August 1976, respectively). They returned valuable samples to Earth and were probably the most productive Soviet lunar missions, but they returned no photographic or cartographic data. They apparently were confined to the opposite side of the nearside equatorial region from the early Luna landers.

In another example of overlap of mission types, photographs were returned by two soft landers, Lunas 17 and 21, which emplaced the roving vehicles Lunakhods 1 and 2. Luna 18 crashed in the sample-return zone (September 1971), and Lunas 19 and 22 (September 1971 and May 1974) were orbiters with television-image transmission systems. In October 1975 the orbiter of Luna 22 was lowered to within 30 km of the surface to take high-resolution pictures. The Lunas 23 and 24 were sample-return missions, but Luna 23 was damaged on landing and therefore unsuccessful. These ended the first round of lunar exploration in August 1976.

APOLLO (1968–72)   The manned Apollo landings were among the most spectacular technological achievements of this century and were the focus of the preceding unmanned programs. For cartographic purposes the value of the orbital flights of Apollo exceeded that of the landing missions. The first seven Apollo missions were unmanned to test spacecraft components and operational procedures. Apollos 8 and 10 were orbital missions (Apollo 9 was an Earth orbiter) that carried hand-held or bracket-mounted Hasselblad cameras and 16-mm "sequence," or variable time-lapse, movie cameras to the Moon (NASA, 1969a, 1971). All subsequent missions carried this equipment and other cameras. Apollos 11, 12, and 14 to 17 were landing missions that employed Hasselblads and 16-mm sequence cameras on the surface and in orbit. In addition, Apollos 11 and 12 employed stereoscopic cameras that took extreme closeup pictures of the surface. Apollos 13 and 14 also carried a special orbital "Hycon" camera designed to take high-resolution pictures of the future Apollo 16 site.

The most systematic and useful orbital coverage was obtained during the last three missions. Apollos 15 to 17 carried sophisticated automatic mapping (metric) and panoramic cameras mounted in the Service Module (Schimerman, 1975; Masursky et al., 1978).

A series of large-scale maps (mostly 1:250,000) called Lunar Topographic Orthophotomaps (LTOs) and Lunar Orthophotomaps (LOs) were constructed along the Apollo ground tracks (Kinsler, 1976). The Apollo data have also been employed to update several LAC charts, to construct similar 1:1,000,000-scale charts of the farside, and to make new charts at 1:5,000,000 scale of the whole surface except, unfortunately, the central nearside.

Lunar exploration has been devoted to landings and selected areas of scientific interest. Unlike most of the other planets, systematic mapping has never been accomplished for the entire lunar surface. The ACIC LAC charts provided regional maps of the nearside and selected areas on the farside. Lunar Orbiter provided photographic coverage of the entire globe, but its geometric integrity is inadequate for global mapping at medium-scale, regional resolution. The mapping cameras carried on the last three Apollo missions provide photogrammetric coverage of a part of the equatorial band comprising approximately 20 percent of the lunar surface. New expeditions have been proposed, including the establishment of one or more inhabited lunar bases from which long exploratory excursions will be made. These cannot take place in the absence of new lunar missions dedicated to finishing the global mapping job and accumulating the data required to make the high-resolution maps that will be required for traverse navigation.

2.2.2. *Mercury*

MARINER 10 (1973–75)   Mariner 10 was launched in November 1973 and was the first spacecraft to visit two planets. One of the goals of the mission was to test the "slingshot" concept, in which the gravity of one planet is used to accelerate a spacecraft toward another, resulting in shorter transit times without extra launch power (Dunne, 1974). This goal was accomplished; not only was the data-gathering efficiency of Mariner 10 dramatically improved, but the technique allowed Voyagers 1 and 2 to fly to both Jupiter and Saturn, and it allowed Voyager 2 to continue on to Uranus and Neptune.

Mariner 10 is in solar orbit. During the Mercury mission, its trajectory was adjusted in such a way that it flew close to Mercury on each of its first three solar orbits (Danielson et al, 1975). The first of these encounters ("Mercury 1") was in March 1974, "Mercury 2" was in September 1974, and "Mercury 3" was in March 1975. Each encounter took place at the same time of the Mercurian day, so that there was no difference in illumination in the pictures taken on any of the encounters. The spacecraft flew past the dark side of Mercury at an altitude of about 700 km on the first encounter in order to evaluate the effects of the planet and its magnetosphere on the solar wind. About thirteen hundred pictures were taken during approach, and nine hundred during departure; there was virtually no rotation of the planet during these sequences, so they covered the illuminated halves of two hemispheres, with no tie between the inbound and the outbound mosaics. On the second encounter, however, the spacecraft was guided past the illuminated hemisphere, passing within 50,000 km at 45° S latitude. About nine hundred frames were taken during this encounter, forming a bridge between the hemispheres. About 45 percent of Mercury is resolved at 1 km per pixel or better on the Mariner 10 images taken during the first two encounters (Davies and Batson, 1975). Three hundred quarter-frames were taken during the third encounter, with pixel dimensions of 100 m. The trajectory of this encounter was almost identical to that of the first. The pictures added to the high-resolution coverage of the surface but did not expand the total aerial coverage.

Stereoscopic convergence exists between first- and second-encounter images (Davies et al., 1978). Preliminary studies indicated that the photogrammetric geometry was unsatisfactory for topographic mapping, however, and the matter was not pursued further. A semicontrolled reference mosaic was made with first-encounter Mariner 10 images, and planimetric shaded relief maps of nine quadrangles were made at 1:5,000,000. A consortium of geologists from government agencies and universities compiled geologic maps on the 1:5,000,000 quadrangles, to be published by USGS. A synoptic map of the illuminated hemisphere was made at 1:15,000,000 in Mercator and Polar Stereographic projections, at 1:10,000,000 in the Lambert Azimuthal Equal-Area projection, and a special shaded relief map of the Caloris Basin was made at 1:5,000,000.

2.2.3. *Venus*

EARLY SPACEFLIGHTS TO VENUS (1962–70)   On August 27, 1962, Mariner 2 began the era of exploration beyond the Moon with spacecraft. Although this spacecraft returned no data suitable for mapmaking, it showed that Venus is blanketed by dense clouds about 25 km thick and that the surface temperatures exceed 425° C. It also showed that the planet has virtually no magnetic field or radiation belts. Mariner 5 flew close to Venus in 1967, but it was not a cartographic mission. During the same year, the Soviet Venera 4 lander entered the Venusian atmosphere but was crushed by atmospheric pressure at an altitude of 34 km. In 1970 Venera 7 survived a landing and for 23 minutes returned pictures of a rock-strewn surface.

MARINER 10 (1974)   Mariner 10 flew past Venus on February 5, 1974, returning pictures of the cloud tops of that enshrouded planet (Murray et al., 1974). Although the pictures showed no surface features and hence could not be used for mapping in the conventional sense, the ones taken in ultraviolet light provided valuable information regarding atmospheric circulation and dynamics.

PIONEER VENUS (1978–?)   In December 1978 a fleet of ten separate spacecraft began an intensive exploration of Venus (Colin, 1980). They included the Pioneer Venus Orbiter, the Pioneer Venus Multiprobe, with a large sounder and three small probes, two Soviet flybys, and two Soviet descent/lander craft designated Venera 11 and 12. The last contributed information about the surface, including images, and the probes made atmospheric measurements. The Pioneer Venus Orbiter carried a radar altimeter and a low-resolution imaging device, from which the first global maps of the Venusian surface were made (Masursky et al., 1980; Pettengill et al., 1980). These maps are at scales of 1:50,000,000. They were produced by making a gridded array of the radii of Venus as measured by the orbiting altimeter and applying digital relief shading techniques to the array. Elevation bands are discriminated by color coding on one version of the map, whereas a second version, designed as a planning chart for future mapping missions, shows contour lines over a subdued shaded relief base.

VENERAS 15 AND 16 (1983–85)   The Soviet Union has devoted much of its planetary exploration efforts to Venus. Their early missions included several successful and nearly successful landers that returned data regarding atmospheric conditions and surface appearance and composition. When the United States abandoned plans for the Venus Orbiter Imaging Radar in 1982, the Soviets initiated a program of topographic and image mapping with synthetic aperture radar.

The Venera 15 and 16 spacecraft (Barsukov et al., 1984) were apparently identical, and placed in nearly identical polar orbits, from which they returned overlapping strips of radar images from the north pole to a latitude of 20° to 25° N. The resolution of these images is about 2 km. Both spacecraft also returned suborbital profiles by radar altimetry. These profiles have closer spacing than did those returned by Pioneer Venus and thus improve the topographic map when combined with Pioneer Venus altimetry. The Soviets have produced and published a set of twenty-seven contoured radar image mosaics, based on Venera 15 and 16 data, at a scale of 1:5,000,000. The United States and the USSR collaborated in a joint mapping project (Basilevsky et al., 1989; USGS, 1989a; 1989b) to produce a 1:15,000,000-scale map of the 25 percent of the northern hemisphere covered by Venera images. This three-sheet map set consists of an airbrushed shaded relief map made with reference to Pioneer Venus/Venera altimetry, Earth-based radar images from Arecibo, and Venera 15/16 radar image mosaics; a contour map on a hypsometric color/shaded relief base; and a radar image mosaic. This set was published by the USGS, along with a geomorphic-geologic map prepared by Soviet authors on the USGS airbrush base map.

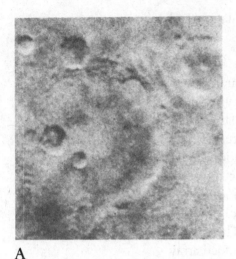

A

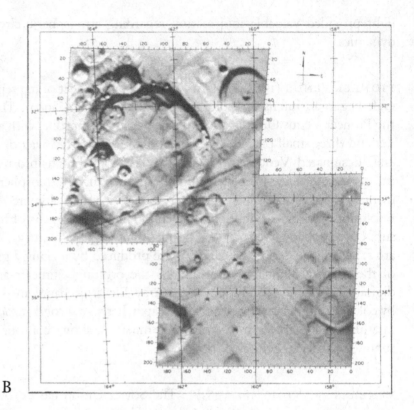

B

Figure 2.21
A. First close-up spacecraft image of Mars, taken from a distance of about 12,000 km. (Mariner 4, 1965, picture 11). B. First map of topographic features on Mars, including part of Sirenum Fossae and the crater "Mariner," made with the airbrush in 1966. (By P. M. Bridges, based on Mariner 4 pictures 11 and 12.)

2.2.3. *Mars*

MARINER 4 (1965)   The first spacecraft to take pictures of Mars was Mariner 4, launched in late 1964. Its sister probe, Mariner 3, failed when the plastic shroud covering the spacecraft could not be ejected, making it impossible to deploy the solar panels. On July 15, 1965, Mariner 4 began transmitting the first of twenty-two pictures of the red planet (Leighton et al., 1967). The images were tiny (200 lines by 200 samples), and the very best ones resolved only about 3 km (the best Earth-based telescopes resolve only about 100 km). Several pictures showed significant, previously unseen surface detail (Figure 2.21A), and an immediate effort was begun to glean every possible bit of information out of each pixel. The new technology of digital image processing was pushed to its limit on each frame. Patricia M. Bridges, who had pioneered the airbrush technique (see above and Chapter 3) on which telescopic lunar mapping was based, was called upon to help make sense of the Mariner 4 pictures; she was later joined by Jay L. Inge, who also became prominent as a lunar and planetary mapper. Although the maps resulting from the Mariner 4 mission were only page-sized illustrations, this project marked the beginning of the collaboration between digital image processing and human photointerpretation that continues to form the basis of planetary mapping (Figure 2.21B; Leighton et al., 1965).

MARINERS 6 AND 7 (1969)   Mariners 6 and 7 passed by Mars in July of 1969 (NASA, 1969b), but their achievement was largely eclipsed by the Apollo 11 landing. Mariner 6 crossed the equatorial zone of Mars in a westerly–easterly direction, whereas Mariner 7 passed from northwest to southeast, providing good coverage of the south polar cap (Leighton et al., 1969a, 1969b). The two spacecraft obtained virtually 100 percent coverage of Mars at resolutions ranging from 50 km per pixel to about 4 km per pixel, as the planet rotated on its axis during their approach. Picture taking began when the spacecraft were nearly 2 million km from Mars, and continued through closest approach at approximately 3,500 km. The "near encounter" pictures, taken between about 10,000 km and closest approach, cover approximately 20 percent of Mars and have average resolutions better than 300 m per pixel. This area proved to be too small to provide a representative sample of Martian geology. A planetwide map of Mars was prepared and published for NASA by the AMS at a scale of 1:25,000,000, based on these pictures. The USGS prepared several experimental quadrangle maps at 1:5,000,000 and 1:1,000,000 in preparation for the Mariner 8 and 9 missions, but these maps were not widely distributed.

MARINER 9 (1971–72)   On November 14, 1971, Mariner 9 became the first spacecraft to orbit a planet beyond the Earth–Moon system (Masursky, 1973). It was launched on May 30, 1971, three weeks after its sister ship, Mariner 8, was lost because of guidance difficulties. During its lifetime of just under one year, Mariner 9 transmitted nearly seventy-three hundred images of Mars, covering nearly 100 percent of the planet with image resolutions of 1 to 3 km per pixel. Its narrow-angle camera provided coverage of approximately 5 percent of Mars, including nearly 70 percent of the south polar region, with image resolutions of 100 to 200 m per pixel.

Mariner 9 data became the basis for the first systematic mapping of an entire planet from space. With the Mariner 9 mission, the USGS in Flagstaff, Arizona, became the primary planetary mapping agency. The planetary cartography group, formed by Raymond M. Batson, was joined by Patricia Bridges and Jay Inge, who in 1970 and 1972, respectively, brought their expertise in airbrush cartography from the ACIC lunar mapping efforts to the USGS planetary work. A planetwide map of Mars was made at 1:25,000,000, and thirty quadrangle maps were made at 1:5,000,000 (Batson, 1976; Batson et al., 1979), along with a selection of larger-scale maps of special interest areas, including maps made in cooperation with the USSR in support of their Mars 4 and 5 missions. Geologic maps were compiled on the 1:5,000,000 quadrangles. Mariner 9 was also used to take pictures of Phobos and Deimos, the two moons of Mars. These were used by Duxbury (1974), Turner (1978), and Stooke and Keller (1989) to make experimental maps of those small, irregularly shaped bodies.

MARS 1 TO 7 (1962–73)   Sputnik 22, Mars 1, and Sputnik 24 were all launched on trajectories for Mars within two weeks of each other during October and November 1962. The launches were detected by radars in the West, and resulted in some consternation in military circles, coming as they did during the Cuban missile crisis. The Sputniks failed early in their missions, but Mars 1 transmitted data until contact was lost, at 106 million km from the Earth. It probably passed within 11,000 km of Mars (Turnill, 1985). Mars 2 reached Mars on November 27, 1971, less than two weeks after Mariner 9 went into Mars orbit. It ejected a lander capsule, but neither the orbiter nor the lander returned cartographic data. Mars 3 was a similar craft, and it arrived five days after Mars 2. Its lander transmitted image data for twenty seconds after it reached the surface. All of Mars was enveloped in a dust storm at this time, however, and the images were never released. The orbiter survived for about three months but evidently returned no useful image data.

The retro-rockets on Mars 4 evidently failed to fire when the spacecraft passed within 2200 km of the planet in February 1974. It returned a strip of images that were incorporated into maps published by the USSR. Mars 5 was successfully placed into orbit, and it returned good pictures of the Nereidum Montes north of Hellas. A landing in that area was intended but was not successful. Mars 6 reached the planet in March 1974; its lander transmitted atmospheric data during descent but fell silent before touchdown. Mars 7 arrived three days before Mars 6 but evidently returned no useful data from either the orbiter or the lander.

VIKINGS 1 AND 2 (1976–82)   In the most comprehensive study of a single planet since the Apollo program, the Viking project consisted of two orbiter and two lander spacecraft (Soffen, 1977). The four spacecraft were launched on two vehicles (each carrying an orbiter and a lander) during the summer of 1975. The Viking 1 Orbiter was inserted into Mars orbit on June 19, 1976; its lander touched down on Mars on July 20 (by chance, the anniversary of both Zond 3 and Apollo 11). Viking 2 entered Mars orbit on August 7, 1976, and the Viking 2 lander touched down on September 3. These spacecraft continued to operate for several years and returned data to provide the foundation for Martian investigations for decades to come.

The Viking Orbiters returned approximately fifty-five thousand images with resolutions ranging from 7 to 1,000 m per pixel. For cartography, one of the most useful sets of Viking Orbiter data consists of approximately five thousand pictures taken with solar elevation angles of about 20 degrees and with resolutions between 100 and 300 m per pixel. These images cover virtually the entire planet, and they are the basis for 140 1:2,000,000-scale controlled photomosaic quadrangles and for a planetwide digital image mosaic. These pictures are also being used to revise the thirty 1:5,000,000 shaded relief quadrangles that were first made from Mariner 9 images. Some 20,000 Viking Orbiter images have resolutions better than 100 m per pixel. Although they are not uniformly distributed over the planet, they are used for special area mapping at scales of 1:500,000 or larger, including controlled photomosaics and geologic maps.

A convergent stereoscopic survey consisting of about one thousand Viking Orbiter pictures was taken from the high-altitude parts of the spacecraft orbits. These images have average resolutions of 800 m per pixel, cover about 70 percent of the planet, and are being used for topographic mapping (see Chapter 6). The remaining 30 percent is covered by images taken for other than cartographic purposes; fortuitously, they allow completion of the mapping, though at reduced accuracy.

Finally, a large number of Viking Orbiter pictures were taken in sets through color filters. These allow multispectral mapping, an aid to geological and geochemical interpretation.

The Viking Landers provided very high resolution (better than 0.5 mm per pixel in some areas) stereoscopic images in both color and black and white of two small areas on Mars. The mapable coverage is less than 1 km^2 at each site, although features visible on the horizon were used by Morris and Jones (1980) to pinpoint the locations of the Viking 1 landing site. A series of forty-two horizon panoramas taken by these spacecraft was published on 90- × 76-cm sheets by the USGS. Maps made with images taken by the landers have not been published individually but have been used as illustrations in scientific reports.

The Viking Landers provided important geodetic benchmarks on Mars, because their coordinates in inertial space could be determined by radio ranging (Michael, 1979). Tying their locations on the Orbiter images with the coordinates determined by radio ranging provided a precise reference frame for the image dataset (Davies and Katayama, 1983; see Chapter 5).

2.2.5. *The Jovian system*

PIONEERS 10 AND 11 (1973–74) Pioneers 10 and 11 visited Jupiter in November 1973 and November 1974, respectively (Fimmel et al., 1977). The scanning photometer on the spinning spacecraft was used to make spectacular color images of the "Red Spot" and the atmospheric bands on Jupiter. Although images of Ganymede and Europa were made with data from these spacecraft, they did not resolve sufficient detail for mapping.

VOYAGERS 1 AND 2 (1979) Voyager 1 was launched on August 20, 1977, and arrived in the Jovian system on March 5, 1979. Voyager 2 was launched on September 5, 1977, and arrived at Jupiter on July 9, 1979. Each spacecraft carried wide- and narrow-angle cameras, which were the primary mapping instruments for the Jovian system. Image resolutions ranged from nominal values of tens of kilometers per pixel taken during the "observatory phase," when the spacecraft approached the system, to a few hundred meters per pixel in two small locations on Io (Smith et al., 1979a, 1979b).

Preliminary airbrush maps were made by USGS cartographers in the mission science area at the Jet Propulsion Laboratory (JPL) as the images arrived from the spacecraft. These were published as uncontrolled maps by the USGS.

The relative motion between the spacecraft and their targets resulted in image smear, which degraded some of the images, but images of sufficiently high quality were taken to make 1:5,000,000-scale maps of Io, one hemisphere of Europa, all of Ganymede (on fifteen quadrangles), and all but the south polar region of Callisto (on fourteen quadrangles) (Batson et al., 1980). The resolution of the images used to make parts of some of these maps was insufficient to justify the 1:5,000,000 scale, but significant areas on each were resolved clearly enough to justify even larger scales.

Geologic mapping is being done on Io, Europa, and Ganymede at 1:5,000,000 scale. Synoptic geologic mapping at 1:15,000,000 is planned for Ganymede and Callisto. Separate synoptic maps of Io and Europa are not being made, because these bodies are small enough to be covered with only three 1:5,000,000 map sheets.

2.2.6. *The Saturnian system*

PIONEER SATURN (1979)   After traversing the Jovian system, Pioneer 11 was redirected to Saturn and renamed "Pioneer Saturn." In September 1979 it became the first spacecraft to pass in the vicinity of the ringed planet. The scanning photometer on the spinning spacecraft was again used to make spectacular color images, this time of Saturn and its rings. Although one new ring and two new satellites were discovered with Pioneer Saturn data (Gehrels et al., 1980), the photometer could not be used to make cartographically meaningful images of the Saturnian satellites.

VOYAGERS 1 AND 2 (1980–81)   On November 12, 1981, Voyager 1 passed through the Saturnian system (Smith et al., 1981). It transmitted moderate-to high-resolution images of the satellites Rhea, Dione, Enceladus, Mimas, Tethys, and Titan (Smith et al., 1982). On August 25, 1981, Voyager 2 made a similar encounter, returning more images of the satellites, including Iapetus. Because of its thick atmosphere, the surface of Titan could not be seen on the Voyager images, so no maps could be made. Preliminary airbrush maps of the remaining six major satellites were made by Bridges and Inge during the encounters and published by the USGS (Batson et al., 1984). Images of eight smaller satellites were also returned. These images had sufficient resolution for morphologic and photometric studies, and Stooke and Keller (1988) prepared experimental maps to demonstrate Stooke's proposed projection system for nonspherical bodies.

2.2.7. *The Uranian system*

VOYAGER 2 (1986)   During a period of only about 10 hours on January 24, 1986, Voyager 2 took high- and medium-resolution pictures of five satellites of Uranus and discovered ten previously unknown ones (Smith, et al., 1986). The Uranian system is tilted on its side, so the sun was almost directly over the south poles of the planet and its satellites. Unlike previous planetary encounters, where satellite rotation resulted in complete photographic coverage, only the southern hemispheres of each of these bodies appear in Voyager 2 images. Image resolution is 4 to 6 km per pixel on Umbriel, Titania, and Oberon, 2 km per pixel at Ariel, and a remarkable 0.3 km per pixel at Miranda, where a mosaic of eight images was made. The rapidly changing position of Voyager relative to Miranda resulted in a strong stereoscopic effect in the parts of the mosaic where individual images overlap, allowing detailed topographic mapping.

Preliminary airbrush maps were again made by Bridges and Inge during the Uranus encounter, which appear in several reports, and final, full-size versions have been published. No attempt has been made to map the newly discovered Uranian satellites.

2.2.8. The Neptunian system

VOYAGER 2 (1989) . In a final, spectacular farewell to the planets of our Solar System, Voyager 2 returned many images of the Neptunian moon Triton, of which eighteen have resolutions of 400 to 1,800 m per pixel and are being used to map about 25 percent of its surface at 1:5,000,000-scale. USGS computer scientists Kathleen Edwards and Eric Eliason have so refined their techniques of digital mosaicking that they were able to prepare, three days after the encounter, a high-quality photomosaic on which the airbrush mapping is based.

Several new moons of Neptune were discovered in Voyager 2 images. One has a diameter of about 400 km and is covered by a single Voyager image with sufficient resolution to permit at least crude mapping. At the time of this writing, that satellite is known only as 1989N1.

2.2.9. Future missions

Several planetary missions planned for the 1990s will contribute to Solar System mapping. These include Mars Observer, the Magellan mission to Venus, the Galileo mission to the Jovian system, and several Soviet Mars missions.

Because the Viking mission was so successful, global and regional mapping of Mars is mostly accomplished. However, the Mars Observer spacecraft will carry a camera system capable of acquiring images at a wide range of scales, including high-resolution (about 1 to 2 m per pixel) frames for small areas. Thus, there will be the opportunity to "fill in" areas not adequately covered during the Viking mission and to obtain data to produce special purpose maps for selected areas, such as future landing sites. The medium-resolution (about 300 m per pixel) mode will produce pole-to-pole, stereoscopic strips with uniform illumination. Assuming sufficiently accurate orbit determinations, it will be theoretically possible to prepare a topographic map of Mars with an overall vertical accuracy of ±500 m.

The Magellan mission to Venus will afford the greatest advance in planetary mapping since the Voyager journey through the outer Solar System. Its principal objective is to obtain radar images for most of the planet at an effective resolution of about 200 m per pixel. Using advanced computer processing and mosaicking techniques, these images will be assembled to produce an atlas of sixty-two 1:5,000,000-scale maps for Venus. Shaded relief versions of these mosaics will also be produced at 1:5,000,000. Because radar images are difficult to use and often contain distortions, the shaded relief maps will probably be more widely used to illustrate the surface of Venus. Plans also call for topographic and geologic maps to be produced in the same format as the radar mosaics and relief maps. In addition, both large- and small-scale maps will be produced for special purposes and for combining with data from previous missions, such as Pioneer Venus and Soviet Venera missions.

The Galileo mission is scheduled to arrive at Jupiter in the mid 1990s to begin a 20-month period of observing the planet and its satellites. Many of the images will be used to complement the Voyager coverage and to image selected areas at high resolution. On its way to Jupiter, the Galileo spacecraft will pass Earth's Moon and, depending upon launch time and trajectory, may provide the opportunity to image areas for which existing coverage is poor, such as the western part of the Orientale basin. In addition, plans call for the spacecraft to fly past one or more asteroids to provide the first closeup views of this class of planetary object.

Thus, most of the planned and proposed missions to be flown throughout the Solar System will carry instruments that will increase the cartographic database for planets, satellites, and other Solar System bodies.

2.3. SUMMARY

Maps of the Moon made with the unaided eye were made as early as 1600. Maps of planets other than the Earth began to appear with the invention of the telescope in 1610, and their production has grown explosively since 1959 (Table 2.1). The first of these were simple sketches. By the mid-1600s lunar maps were made that included nomenclature schemes. Some elements of the early nomenclature became the basis for modern planetary nomenclature. Quantitative mapping with geometric controls began in the mid-1700s.

The topography of the Moon is well resolved with even primitive telescopes, providing observers with opportunities to make detailed maps showing mountains and craters, as well as the light- and dark-albedo markings that are visible with the naked eye. Of the other (solid surface) planets, only Mercury and Mars are resolved as more than mere points of light, even with the best telescopes, so that only albedo markings can be seen on these bodies. This

Table 2.1. *Milestones in planetary mapping.*

Author or Platform	Date	Instrument	Cartographic Result
Gilbert	1600	Naked eye	Full-Moon sketch
Galileo	1610	Telescope	Sketches of phase
Langrenus	1645	Telescope	Large full-Moon shaded relief map with albedo and nomenclature scheme
Grimaldi, Riccioli	1651	Telescope	Core of present nomenclature
Huygens	1659	Telescope	Sketch map of Mars; Syrtis Major identifiable
Mayer	1749	Telescope with measuring reticule	Lunar control net
Lohrmann	1824	Telescope	Comprehensive and accurate lunar mapping
Beer, Mädler	1838	Telescope	Published Mars map; 0° meridian defined
Schiaparelli	1878	Telescope	Mars map with nomenclature forms the basis of current system
ACIC/Lowell	1959	Telescope, telescopic photos	Lunar 1:1,000,000 mapping
Luna 3	1959	Television camera	View of lunar farside
Kuiper, Arthur, Whitaker	1960	Telescopic photos	Rectified lunar atlas
Shoemaker, Hackman	1962	Telescope, telescopic photos	Defined and illustrated principles of lunar geologic mapping
Goldstein	1963	Earth-based radar	Reflectivity profiles of Mars and Venus
Ranger	1964	Television camera	High-resolution maps of lunar surface
Mariner 4	1965	Television camera	Images of Mars topography
Lunar Orbiter	1966	Electronic film scan	Systematic high-resolution coverage of Moon
Apollo 11	1969	Retro-reflector	Lunar geodesy by laser-ranging from Earth
Apollo 15	1971	Returned film	Lunar topographic mapping by photogrammetry
Mariner 9	1972	Television camera	Systematic mapping of Mars
Mariner 10	1974	Television camera	Systematic survey of Mercury
Campbell	1975	Earth-based radar	Radar-image mosaics of Venus
Viking	1976	Spacecraft radio transmitter	Geodesy from Martian surface by radio-tracking
Pioneer Venus	1978	Radar altimetry	Global altimetric survey of 70% of Venus' surface
Venera 15–16	1983	Synthetic-aperture radar	SAR image survey of northern hemisphere of Venus

did not deter early observers from sketching their observations in map form; sketches of Mars showing features recognizable today were made as early as 1636. Meaningful observations of Mercury were made in the 1880s.

The advent of radar astronomy in the late 1950s provided mappers with quantitative, if sometimes ambiguous, information about the topography of the Mercurian, Venusian, and Martian surfaces. Topographic profiles in the equatorial regions of Mars and Mercury were derived from this technology, and the first images of the cloud-veiled surface of Venus were made by radar.

In this century planetary cartography did not become a fully respectable endeavor until the advent of spaceflight. Accurate maps were then needed prior to spacecraft missions for preflight guidance and planning and afterward to arrange and categorize spatial data successfully returned to Earth. Maps were made by systematic telescopic observation and photography and with the film or television signals returned to Earth by cameras mounted on spacecraft. More than a thousand photomosaics, shaded relief maps, topographic maps, and geologic maps of twenty planets and satellites have been published, nearly all of them since 1959.

2.4. REFERENCES

Antoniadi, E. M. 1930, *La planète Mars*. Paris: Libraire Scientifique Herman, pl. II, III, IV, V.

Antoniadi, E. M. 1933. La planète Mercure. *J. Royal Astr. Soc. Canada*, 27(10): 403–10.

Arthur, D. W. G., and Agnieray, A. P. 1964. *Lunar Designations and Positions, Quadrants I, II, and III*. Univ. of Arizona Lunar Planet. Lab. (3 charts).

Arthur, D. W. G., Agnieray, A. P., Horvath, R. A., et al. 1963a. *The System of Lunar Craters, Quadrant I*. Commun. no. 30, Univ. of Arizona Lunar Planet. Lab.

Arthur, D. W. G., Agnieray, A. P., Horvath, R. A., et al. 1963b. *The System of Lunar Craters, Quadrant II*. Commun. no. 40, Univ. of Arizona Lunar Planet. Lab.

Arthur, D. W. G., Agnieray, A. P., Pellicori, R. H., et al. 1965. *The System of Lunar Craters, Quadrant III*. Commun. no. 50, Univ. of Arizona Lunar Planet. Lab.

Arthur, D. W. G., and Pellicori, R. H. 1965. *Lunar Designations and Positions, Quadrant IV*. Univ. of Arizona Lunar Planet. Lab. (chart).

Arthur, D. W. G., Pellicori, R. H., and Wood, C. A. 1966. *The System of Lunar Craters, Quadrant IV*. Commun. no. 70, Univ. of Arizona Lunar Planet. Lab.

Arthur, D. W. G., and Whitaker, E. A. 1960. *Orthographic Atlas of the Moon – Supplement 1 to the Photographic Lunar Atlas*. Tucson: Univ. of Arizona Press.

Barabashov, N. P., Mikhailov, A. A., and Lipsky, Yu. N. 1961. *An Atlas of the Moon's Far Side: The Lunik III Reconnaissance*. New York and Cambridge, Mass.: Interscience. (Originally published in 1960 in Russian. Moscow: Academiia Nark.)

Barsukov, V. L., Bazilevsky, A. T., Kuzmin, R. O., et al. 1984. Geologiya Venery rezultatum analiza radiolokatsionnykh izobrazhenii, poluchennykh AMC Venera-16 i Venera-16 (predvaritelnye dannye). *Geokhimiya* 12: 1811–20.

Basilevsky, A. T., Burba, G. A., and Batson, R. M. 1989. Maps of part of the Venus northern hemisphere: A joint U.S./U.S.S.R. mapping project. In *Abstracts of Papers Submitted to the Twentieth Lunar and Planetary Science Conference, Houston, March 13–17, 1989*, Lunar and Planetary Institute, pp. 46–7.

Batson, R. M. 1976. Cartography of Mars: 1975. *The American Cartographer* 3 (1): 57–63.

Batson, R. M., Bridges, P. M., and Inge, J. L. 1979. *Atlas of Mars: The 1:5,000,000 Map Series*. National Aeronautics and Space Administration Spec. Pub. 438.

Batson, R. M., Bridges, P. M., Inge, J. L., et al. 1980. Mapping the Galilean satellites of Jupiter with Voyager data. *Photogrammetric Engineering and Remote Sensing* 46 (10): 1303–12.

Batson, R. M., Jordan, Raymond, and Larson, K. B. 1974. *Atlas of Surveyor 5 Television*

Data. National Aeronautics and Space Administration Spec. Pub. 341.

Batson, R. M., Lee, E. M., Mullins, K. F., et al. 1984. *Voyager 1 and 2 Atlas of Six Saturnian Satellites*. National Aeronautics and Space Administration Spec. Pub. 474.

Beer, W., and Mädler, J. H. 1838. *Der Mond nach seinen komischen und individuellen Verhältnissen, oder Allgemeine vergleichende Selenographie*. Berlin: Simon Schropp.

Blagg, M., and Müller, K. 1935. *Named Lunar Formations*. Vol. 2, *Maps*. London: Percy Lund, Humphries.

Campbell, D. B., and Burns, B. A. 1980. Earth-based radar imagery of Venus. *J. Geophys. Res.* 85 (A13): 8271–81.

Colin, Lawrence. 1980. The Pioneer Venus program. *J. Geophys. Res.* 85 (A13): 7575–98.

Cruikshank, D. P., and Chapman, C. R. 1967. Mercury's rotation and visual observations. *Sky and Telescope* 34: 24–6.

Danielson, G. E., Jr., Klaasen, K. P., and Anderson, J. L. 1975. Acquisition and description of Mariner 10 television science data at Mercury. *J. Geophys. Res.* 80 (17): 2357–93.

Davies, M. E., and Batson, R. M. 1975. Surface coordinates and cartography of Mercury. *J. Geophys. Res.* 80 (17): 2417–30.

Davies, M. E., Dwornik, S. E., Gault, D. E., and Strom, R. G. 1978. *Atlas of Mercury*. National Aeronautics and Space Administration Spec. Pub. 423.

Davies, M. E., and Katayama, F. Y. 1983. The 1982 control network of Mars. *J. Geophys. Res.* 88 (B9): 7403–4.

Dunne, J. A. 1974. Mariner 10 Mercury encounter. *Science* 185: 141–2.

Duxbury, T. C. 1974. Phobos: Control network analysis. *Icarus* 23: 290–9.

Eggleton, R. E. 1965. Geologic map of the Riphaeus Mountains region of the Moon. USGS Miscellaneous Investigations Series Map I–458 (LAC 76; scale 1:1,000,000).

Fimmel, R. O., Swindell, W., and Burgess, E. 1977. *Pioneer Odyssey, Encounter with a Giant*. National Aeronautics and Space Administration Spec. Pub. 349.

Franz, J. 1913. Die Randlandschafen des Mondes. *Nova Acta, Abh. der Kaiserl. Leop.-Carol. Deutschen Akad. der Naturforscher* 99 (1): 1–96.

Gehrels, T., Baker, L. R., Beshoye, E., et al. 1980. Imaging photopolarimeter on Pioneer Saturn. *Science* 207: 434–9.

Goldstein, R. M., Green, R. R., and Rumsey, H. C. 1978. Venus radar brightness and altitude images. *Icarus* 36: 334–52.

Goodacre, W. 1931. *The Moon*: Bournemouth: Pardy & Son.

Hackman, R. J. 1961. Photointerpretation of the lunar surface. *Photogrammetric Engineering* (June 1961): 377–86.

Hackman, R. J., and Mason, A. C. 1961. Engineer special study of the surface of the Moon. U.S. Geological Survey Misc. Inv. Series Map I–351 (scale 1:3,800,000).

Hall, R. C. 1977. *Lunar Impact: A History of Project Ranger*. National Aeronautics and Space Administration Spec. Pub. 4210.

Heacock, R. L., Kuiper, G. P., Shoemaker, E. M., Urey, H. C., and Whitaker, E. A. 1965. *Ranger VII, Part II*. National Aeronautics and Space Administration – Jet Propulsion Lab. Tech. Rpt. 32–700.

Heacock, R. L., Kuiper, G. P., Shoemaker, E. M., et al. 1966. *Rangers VIII and IX, Part II*. National Aeronautics and Space Administration – Jet Propulsion Lab. Tech. Rpt. 32–800.

Hevelius, J. 1647. *Selenographia sire lunae descriptio*. Danzig.

Huygens, C. 1925. *Oeuvres complètes de Christiaan Huygens*. Vol. 15. The Hague: M. Nijhof, p. 64. Reprinted 1967, Amsterdam: Swets & Zeitlinger.

IAU. 1960. *Transactions X*. Dordrecht: D. Reidel, p. 263.

Kinsler, D. C. 1976. New lunar cartographic products. *Proc. Lunar Sci. Conf. 7* 3: i–x.

Kopal, Zdenek, and Carder, Robert W. 1974. *Mapping of the Moon*. Dordrecht/Boston: D. Reidel.

Kuiper, G. P., Arthur, D. W. G., Moore, P., et al. 1960. *Photographic Lunar Atlas*. Chicago: Univ. of Chicago Press.

Kuiper, G. P., Whitaker, E. A., Strom, R. G., et al. 1967. *Consolidated Lunar Atlas – Supplements 3 and 4 to the USAF Photographic Atlas*. Univ. of Arizona Lunar Planet. Lab., Contrib. no. 4.

Leighton, R. B., Horowitz, N. H., Murray,

B. C., et al. 1969a. Mariner 6 television pictures: First Report. *Science* 165: 684–90.

Leighton, R. B., Horowitz, N. H., Murray, B. C., et al. 1969b. Mariner 7 television pictures: First Report. *Science* 165: 787–95.

Leighton, R. B., Murray, B. C., Sharp, R. P., et al. 1965. Mariner IV photography of Mars: Initial results. *Science* 149: 627–630.

Leighton, Robert B., Murray, Bruce C., Sharp, Robert P., et al. 1967. *Mariner IV Pictures of Mars*. Jet Propulsion Laboratory Technical Report 32–884.

Lipsky, Yu. N. 1965. Zond–3 photographs of the Moon's farside. *Sky and Telescope* 30: 338–41.

Lipsky, Yu. N. 1966. What Luna 9 told us about the Moon. *Sky and Telescope* 32: 257–60.

Lipsky, Yu. N., ed. 1969. *Atlas of the Reverse Side of the Moon, Part II*. NASA Tech. Transl. TT F–514. (Atlas obratnoy storony luny, chast II, Moscow, Nanka, 1967.)

Loewy, M., and Puiseux, P. 1896–1909. *Atlas Photographique de la Lune*. Paris, l'Observatoire de Paris.

Lohrmann, W. G. 1824. *Topographie der Sichtbaren Mondoberfläche*. Dresden and Leipzig: J. F. Hartknach.

Lowell, P. 1923. In *Splendour of the Heavens*, vol. 1., T. E. R. Phillips and W. H. Steavenson, eds. London: Hutchinson.

Lucchitta, B. K. 1978. *Geologic Map of the North Side of the Moon*. USGS Map I–1062 (scale 1:5,000,000).

Marshall, C. H. 1963. *Geologic Map and Sections of the Letronne Region of the Moon*. USGS Map I–385 (LAC 75; scale 1:1,000,000).

Mason, A. C., and Hackman, R. J. 1962. Photogeologic study of the Moon. In Kopal, Z., and Mikhailov, Z. K., eds., *The Moon*. London: Academic Press, pp. 301–15.

Masursky, Harold. 1973. An overview of geological results from Mariner 9. *J. Geophys. Res.* 78 (20): 4009–30.

Masursky, Harold, Colton, G. W., and El-Baz, Farouk, eds. 1978. *Apollo over the Moon: A View from Orbit*. NASA SP–362.

Masursky, Harold, Eliason, Eric, Ford, Peter G., et al. 1980. Pioneer Venus radar results: Geology from images and altimetry. *Geology J. Geophys. Res.* 85 (A13): 8232–60.

Lichtenberg, G. C. 1775. *Opera Inedita*. Göttingen: Publisher unknown.

Michael, W. H., Jr. 1979. Viking Lander tracking contributions to Mars mapping. *The Moon and Planets* 20: 149–52.

Morris, E. C., and Jones, K. L. 1980. Viking 1 lander on the surface of Mars: Revised location. *Icarus* 44: 217–22.

Murray, B. C., Belton, M. J. S., Danielson, G. E., et al. 1974. Venus: Atmospheric motion and structure from Mariner 10 pictures. *Science* 183: 1307–15.

Mutch, T. A. 1970. *Geology of the Moon – A Stratigraphic View*. Princeton, N.J.: Princeton Univ. Press.

NASA. 1969a. *Analysis of Apollo 8 Photographs and Visual Observations*. NASA SP–201.

NASA. 1969b. *Mariner–Mars 1969: A Preliminary Report*. National Aeronautics and Space Administration Spec. Report 225.

NASA. 1971. *Analysis of Apollo 10 Photographs and Visual Observations*. NASA SP–232.

Neison, E. 1876. *The Moon, and the Condition and Configuration of Its Surface*. London: Longmans, Green.

Pettengill, Gordon H., Eliason, Eric, Ford, Peter G., et al. 1980. Pioneer Venus radar results: Altimetry and surface properties. *J. Geophys. Res.* 85 (A13): 8261–70.

Proctor, R. A. 1892. *Astronomy Old and New*: London and New York: Longmans, Green.

Riccioli, G. B. 1651. *Almagestum Novum*. Bologna.

Saunder, S. A. 1911. Determination of positions and the measurement of lunar photographs. *Memoirs Royal Astron. Soc.* 60: 1–81.

Schiaparelli, G. V. 1881. *Mappa Areographica*. In *Atti della R. Accademia dei Lincei*. Ser. IIIa, vol. 10, pl. 3.

Schimerman, L. A., ed. 1975. *Lunar Cartographic Dossier*. St. Louis: Defense Mapping Agency, Aerospace Center.

Schmidt, J. F. J. 1878. *Mondcharte von Wilhelm Gotthelf Lohrmann*. Leipzig: J. A. Barth.

Schröter, J. H. 1791. *Selenotopographishe Fragmente*, vol. 1. Lilienthal: C. G. Fleckeisen.

Scott, D. H., McCauley, J. F., and West, M. N. 1977. *Geologic Map of the West Side of the Moon*. USGS Map I–1034 (scale 1:5,000,000).

Shoemaker, E. M. 1962a. Interpretation of lunar craters. In Kopal, Z., ed., *Physics and Astronomy of the Moon*. New York, Academic Press, pp. 283–359.

Shoemaker, E. M. 1962b. Exploration of the Moon's surface. *American Scientist* 50: 99–130.

Shoemaker, E. M. 1981. Lunar geology. In Hanle, P. A., and Chamberlain, V. D., eds, *Space Science Comes of Age*. Washington, D.C.: Smithsonian Institution, pp. 51–5.

Shoemaker, Eugene M., Batson, Raymond M., and Larson, Kathleen B. 1966. An appreciation of the Luna 9 pictures. *Astronautics and Aeronautics* 4 (5): 40–50.

Shoemaker, E. M., and Hackman, R. J. 1962. Stratigraphic basis for a lunar time scale. In Kopal, Z., and Miklhalov, S. K., eds., *The Moon*. London: Academic Press, pp. 289–300.

Shoemaker, E. M., Morris, E. C., Batson, R. M., et al. 1969. Television Observations from Surveyor. In *Surveyor Program Results*. NASA SP–184, pp. 19–128.

Smith, B. A., Soderblom, L. A., Batson, R. M., et al. 1982. A new look at the Saturn system: The Voyager 2 images. *Science* 215: 504–37.

Smith, B. A., Soderblom, L. A., Beebe, R., et al. 1979a. The Galilean satellites and Jupiter: Voyager 2 imaging science results. *Science* 206: 925–50.

Smith, B. A., Soderblom, L. A., Beebe, R., et al. 1981. Encounter with Saturn: Voyager 1 imaging science results. *Science* 212: 163–91.

Smith, B. A., Soderblom, L. A., Beebe, R., et al. 1986. Voyager 2 in the uranian system: Imaging science results. *Science* 233: 43–64.

Smith, B. A., Soderblom, L. A., Johnson, T. V., et al. 1979b. The Jupiter system through the eyes of Voyager 1. *Science* 204: 951–72.

Soffen, Gerald A. 1977. The Viking project. *J. Geophys. Res.* 82 (28): 3959–70.

St. Clair, J. H., Carder, R. W., and Schimerman, L. A. 1979. United States lunar mapping – a basis for and result of Project Apollo. *The Moon and the Planets* 20: 127–48.

Stooke, Philip J., and Keller, C. Peter. 1989.

Map projections for non-spherical worlds: The variable-radius map projections. *Cartographica* [in press].

Stuart-Alexander, D. E. 1978. *Geologic Map of the Central Far Side of the Moon*. USGS Map I–1047 (scale 1:5,000,000).

Trask, N. J. 1972. *The Contribution of Ranger Photographs to Understanding the Geology of the Moon*. USGS Prof. Paper 599-J.

Turner, R. 1978. A model of Phobos. *Icarus* 33: 116–40.

Turnill, Reginald. 1985. *Jane's Spaceflight Directory*. London: Jane's.

U.S. Geological Survey. 1989. Topographic map of part of the northern hemisphere of Venus. USGS Miscellaneous Investigations Series Map I–2041 (scale 1:15,000,000).

Wilhelms, D. E. 1970. *Summary of Lunar Stratigraphy – Telescopic Observations*. USGS Prof. Paper 599-F.

Wilhelms, D. E., and El-Baz, Farouk. 1977. *Geologic Map of the East Side of the Moon*. USGS Map I–946 (scale 1:5,000,000).

Wilhelms, D. E., Howard, K. A., and Wilshire, H. G. 1979. *Geologic Map of the South Side of the Moon*. USGS Map I–1162 (scale 1:5,000,000).

Wilhelms, D. E., and McCauley, J. F. 1971. *Geologic Map of the Near Side of the Moon*. USGS Map I–703 (scale 1:5,000,000).

Wilhelms, D. E., Trask, N. J., and Keith, J. A. 1965. Compilation of geology in the lunar equatorial belt. *Astrogeologic Studies Annual Progress Report, July 1964–July 1965*, map supplement (scale 1:5,000,000) (USGS open-file report).

Wilkins, H. P., and Moore, P. A. 1961. *The Moon*. New York: Macmillan.

Whitaker, E. A. 1963. Evaluation of the Soviet photographs of the Moon's far side. In Middlehurst, B. M., and Kuiper, G. P., eds., *The Moon, Meteorites, and Comets*. Chicago: Chicago Univ. Press, pp. 123–8.

Whitaker, E. A., Kuiper, G. P., Hartmann, W. K., and Spradley, L. H. 1963. *Rectified Lunar Atlas: Supplement Number Two to the Photographic Lunar Atlas*. Tucson: Univ. of Arizona Press.

Woods, David. 1981. Probes to the Moon. In *The Illustrated Encyclopedia of Space Technology*. New York: Harmony, pp. 128–37.

3

Cartography

RAYMOND M. BATSON

3.1. MAP DESIGN

3.1.1. *General considerations*

Maps of the Earth and planets are made by projecting the surfaces of spheroids onto flat sheets of paper. A consequence of the projection process is the distortion of planetary features on the resulting maps. These distortions can be reduced to a minimum and controlled mathematically by judicious selection of map projections, but the discrepancy between map scale and true scale can never be eliminated. Scale distortions are most obvious on small-scale maps (maps that cover large areas). For example, the scale of a Mercator projection is twice as large at the sixtieth parallel as it is at the equator.

Maps are needed to show shapes, dimensions, and areas, but no one map projection can achieve all of these goals well. Conformal projections like the Mercator, Transverse Mercator, Lambert Conformal Conic, and Polar Stereographic retain the true shapes of small landforms and are favored by scientists wishing to recognize and interpret phenomena characterized by shape. Equal-area projections, on the other hand, are useful in evaluating the distributions of surface features, such as craters, especially at global scale. Conformal projections have been used by planetary scientists for many years, but equal-area maps are now being produced as well.

Reasonable characterizations of entire planets at page size pose unique cartographic problems. The significant properties of most landforms do not show at such small scales and must be exaggerated slightly, whereas the least significant must be subdued to avoid clutter. Lambert Azimuthal Equal-Area projections are useful for this purpose. This projection has the advantage that any closed figure (a circle or square, for example) of given dimensions covers exactly the same area on the planet, regardless of location. However, shapes of features are not preserved. Craters appear as ellipses at the edge of the map and as circles in the center, but this apparent foreshortening is not nearly as pronounced as it is on orthographic projections, which produce a similar effect.

The mapping of small, irregular satellites such as Phobos pose unique problems. These are primitive objects, which contain valuable clues regarding the origin of the Solar System. They do not lend themselves to mapping by conventional methods because mathematical projections of their surfaces fail to convey an accurate visual impression of landforms and because map scale varies with the shapes of the bodies, making feature measurements difficult. Maps of these bodies are therefore compiled and stored in digital form, in such a way that visual representations of the maps can be made in any desired geometric form. For example, Figure A-I.1 (Appendix I) has been used to project three-dimensional digital mosaics of spacecraft images to six orthogonal orthographic views. Stooke and Keller (1988) have proposed a special projection for irregularly shaped objects and have compiled several maps to illustrate their concept.

Good cartographic design requires the selection of optimum projection parameters to minimize distortion for each map series. The standard parallels of a band of Lambert Conformal Conic projections, for example, are generally selected to minimize scale distortions within that band alone. A common practice is to locate the standard parallels one-sixth of the width of the band from its northern and southern boundaries. For example, if a band of conic projections lies between the thirtieth and the sixtieth parallels, the optimum standard parallels will be at latitudes 35° and 55°. Two sets of standard parallels would therefore be indicated when two bands of Lambert conic projections are used; this practice is not always followed in planetary cartography because map users frequently need to make mosaics of maps to cover very large areas and to tie work done at one scale to work done at another. Maps made on different standard parallels will not match at their common borders. Several planetary map series, therefore, are simply rescaled subdivisions of others and have the same projection geometry as the parent series. In such cases, two bands of Lambert conic projections will be based on common standard parallels, and only one of these parallels will lie within the boundaries of any one map sheet. Similarly, the latitude of true scale may lie outside the boundary of some of the Mercator map sheets when they are subdivisions of smaller-scale Mercator map sheets.

Planets must be divided into projection zones bounded by meridians or parallels for systematic mapping. Different projections or projection parameters are used in each of these zones. The Mercator projection is used in equatorial latitudes; the Lambert Conformal Conic (with two standard parallels) in the intermediate latitudes of Mercury, Venus, Mars, Callisto, and Ganymede; and the Polar Stereographic in polar latitudes.

Adjacent projections usually share the same scale at their common boundaries, but map details will match only at points of tangency. For example, whereas the thirtieth parallel of a Mercator projection is a straight line, this parallel is a segment of a circle on a Lambert Conic. Similarly, a Transverse Mercator projection centered on one meridian matches one centered on another meridian only at a single tangent point on common boundary meridians because these meridians curve in opposite directions.

Imprecise ground control and paper shrinkage commonly cause more significant distortions than does projection on large-scale maps. Most planetary maps have small scales by terrestrial standards because they are made from global and near-global views. Only rarely does a dataset justify mapping at scales larger than 1:1,000,000. The characteristics of projections used for base maps therefore tend to be more important to scientists interested in global studies than to those interested in small areas. As exploration has progressed beyond the reconnaissance stage, however, and successively higher resolution images have become available, new map series at larger scales have been designed. Maps that have scales as large as or larger than 1:500,000 are now possible for much of the Moon and Mars.

3.1.2. Scales

Scales are exact only at particular points or along particular lines in any map projection. Scale, when used to designate a series (e.g., 1:25,000,000), is the scale at some specified location, such as the equator or a standard parallel. Scale factors must be used to calculate true scales at other locations. A scale factor is the number that, when multiplied by the nominal scale of a map, gives the true scale at some latitude (or longitude or point, on some projections). For example, the Mercator projection with a nominal (i.e., equatorial) scale of 1:25,000,000 has a scale factor of 1.7883 at 56°. Its true scale at that latitude, therefore, is (1.7883) $(1/25,000,000) = 1/13,979,757$. Similarly, the scale factor at the pole for the Mars 1:25,000,000 map is 1.8589, so the scale at the pole is (1.8589) $(1/25,000,000) = 1/13,449,000$.

Scale factors at critical locations are given in Appendix I. These critical locations are latitudes of true scale or the latitude boundaries between projections.

3.1.3. Projections

Map projections can be visualized as planes, cones, or cylinders tangent to or intersecting a sphere (see Appendix I; Snyder, 1982, 1987). Features on the globe are projected onto the planes, cones, or cylinders according to some mathematically defined system. The Polar Stereographic is the simplest conformal projection used in planetary cartography, and it can be constructed graphically. Although it is sometimes useful to think of Transverse and normal Mercator projections as cylinders and of Lambert Conformal Conic projections as cones, these are mathematically modified to achieve the conformal condition and are not graphical projections. Standard parallels, central meridians, and projection centers define the location of true or constant scale on map projections. These are the lines or points at which a projection intersects or lies tangent to a globe.

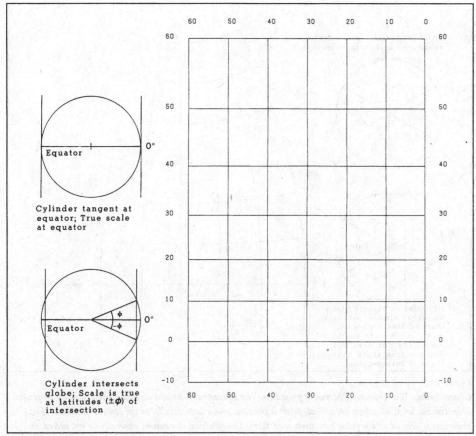

Figure 3.1
Characteristics of a Mercator projection. Meridians are equally spaced, straight lines that are perpendicular to parallels and that have progressively greater spacing away from the equator. The projection cannot be used for maps of polar regions.

MERCATOR The Mercator projection (Figure 3.1) is used for planetary mapping at all scales. The normal Mercator projection is tangent to a globe at the equator or intersects it at some pair of latitudes north and south of the equator. It provides a convenient rectangular format for all but the polar regions of a planet. The fact that it is conformal allows easy recognition of features appearing on spacecraft images. The primary disadvantage of the Mercator projection is that the scale changes vary rapidly with latitude; map portrayals of features are much smaller in the equatorial regions than in higher or lower latitudes. The Mercator projection is undefined at the poles of a planet, and other projections, such as the Polar Stereographic, must be employed.

TRANSVERSE MERCATOR The Transverse Mercator projection (Figure 3.2) is simply a Mercator projection turned on its side. Mathematically, scale varies with distance from the central meridian in the same way as on the normal

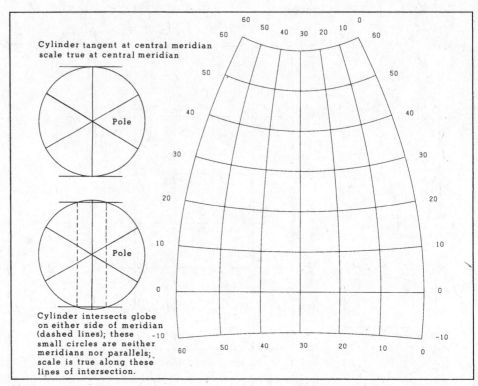

Figure 3.2
Characteristics of a Transverse Mercator projection. The equator and central meridian are mutually perpendicular straight lines; all others are curved. Scale is constant along lines parallel to the central meridian. The projection is defined at the poles, but rarely used there. Longitudinal dimensions generally do not exceed 20 degrees because of rapid scale change beyond that limit.

Mercator. Because of the difference in definition between meridians and parallels, however, scale change with latitude and longitude is mathematically more complicated than it is on the normal Mercator projection. The Transverse Mercator projection is used for planetary mapping at scales larger than 1:1,000,000. The central meridian of a Transverse Mercator projection is a straight line perpendicular to the equator, which is also a straight line. All other meridians are complex curves with their concave sides toward the central meridian, and parallels are complex curves with their convex sides toward the equator.

LAMBERT CONFORMAL CONIC   The Lambert Conformal Conic (Figure 3.3) can be visualized as a cone tangent to or intersecting a globe. The apex of the cone is pierced by the spin axis of the planet. The apex angle depends on the latitude at which the cone is tangent to the globe or the latitudes at which the cone intersects the globe. These latitudes are called standard parallels. The standard parallels are usually selected as the latitudes of true scale. Scale is constant at any given latitude, but it is larger between the standard parallels and smaller outside the zone bounded by the standard parallels.

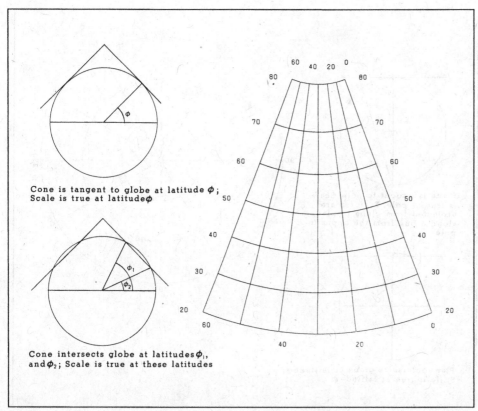

Cone is tangent to globe at latitude φ;
Scale is true at latitudeφ

Cone intersects globe at latitudes φ₁,
and φ₂; Scale is true at these latitudes

Figure 3.3
Characteristics of a Lambert Conformal Conic projection. Meridians are radiating straight lines; parallels are segments of concentric circles with radii that vary with distance from the standard parallels.

POLAR STEREOGRAPHIC The Polar Stereographic projection (Figure 3.4) is a plane that is tangent to the globe at either pole or that intersects it at some latitude. Sometimes referred to as the latitude of true scale, it is analogous to the standard parallel of the Lambert Conformal Conic projection. Oblique stereographic projections, in which the plane is tangent somewhere besides the pole, are used on rare occasions for special-purpose maps of the planets.

LAMBERT AZIMUTHAL EQUAL-AREA The Lambert Azimuthal Equal-Area projection (Figure 3.5) can be visualized as a plane tangent to the globe, but there is no simple graphical analog to explain the mathematical projection of meridians and parallels from the globe to the plane. A crater in the center of a planet's hemisphere is round, but an identical crater located near the edge of the map is elliptical. Both mapped images have the same area, even though their shapes are different.

The true azimuth from the center of this projection to any point in the projection can be determined directly by drawing a straight line radial from the center of the projection.

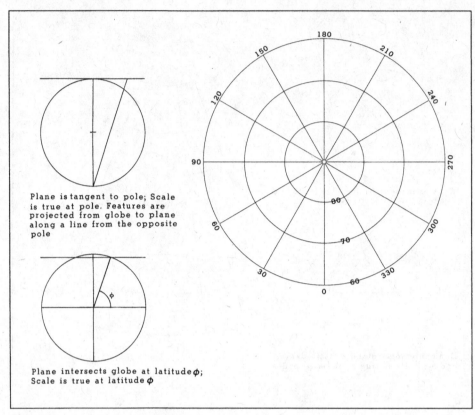

Plane is tangent to pole; Scale is true at pole. Features are projected from globe to plane along a line from the opposite pole

Plane intersects globe at latitude φ; Scale is true at latitude φ

Figure 3.4

Characteristics of a Polar Stereographic projection. Meridians are radiating straight lines; parallels are concentric circles with radii that vary with distance from the pole.

3.1.4. *Map series*

A map series is a set of maps of a specific planet that have a specified scale, projection scheme, and map type (planimetric, topographic, geologic, etc.). Thus, planetary map series are designed to specific requirements:

1. To produce small-scale and synoptic maps showing entire planets on single sheets for data indexing and planning and as bases for global geologic maps
2. To produce a series of regional maps in which planets are segmented into quadrangles as necessary, depending on resolution of available data
3. To produce standard map scales for similar map series, regardless of planet, for comparative studies.

Map series designs are based on the size of a planet, desired map scales, and resolution of available data. The planet usually must be divided into segments (i.e., quadrangles). The more segments and the smaller the area covered, the better each can be made to fit on a flat sheet of paper. However, where data resolution is low and a planet is small, as in the Voyager images

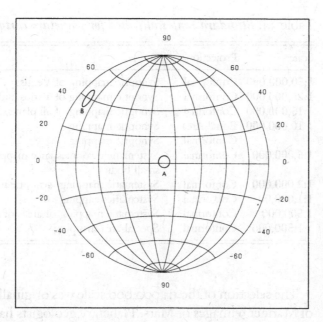

*Figure 3.5
Characteristics of a Lambert
Azimuthal Equal-Area projection. The equator and central
meridian (if drawn) are mutually perpendicular straight
lines. The area of a map of a
hemisphere $(\pi y)^2$ equals the area
of a hemisphere of a scaled globe
$(4\pi y)^2$.*

of the Saturnian satellites, extensive subdivision is not useful because it results
in map sheets that are absurdly small or in scales that are too large to be
supported by available data. The scale distortions inherent in synoptic Mercator, Polar Stereographic, or Lambert Azimuthal Equal-Area projections
must therefore be accepted for this kind of mapping.

Maps of entire planets are made on two conformal projections, Mercator
and Polar Stereographic. The use of similar scales facilitates comparison of
landforms and other features appearing on the maps, but it is not always
feasible because of the wide range in sizes of planets being mapped. In some
cases the same map compilation is published at different scales and in slightly
different formats (see Table A-I.b). For example, the Venus map compiled at
1:15,000,000 on the eight-sheet, synoptic format (Figure A-I.4) is reproduced
on three sheets at the reduced scales of 1:25,000,000 (Figure A-I.3) and on
one sheet at 1:50,000,000 (Figure A-I.2).

Small-scale synoptic maps are composites of all available map data for a
given planet. Interpretation and generalization of landforms is frequently necessary when making these maps, so that details do not obliterate or obscure
regional or global trends.

For more detailed regional work, the planetary scientist usually prefers to
work with a minimum number of map sheets, each of a convenient size and
each presenting a high-resolution picture of a planetary surface. Although
preferences tend to be somewhat mutually exclusive, a scale of 1:5,000,000
has been selected as the standard scale for systematic regional reconnaissance
mapping of the planets. This scale permits representation of regional geo-
morphology and is commensurate with the resolution of most image datasets
from planetary spacecraft. Even the largest terrestrial planets (Earth and Venus) can be mapped at 1:5,000,000 on fewer than 100 sheets.

Table 3.1. *Standard map scales used for planetary cartography.*

Scale	Projection	Use
1:50,000,000	Conformal	Synoptic mapping of Venus
1:25,000,000	Conformal	Synoptic mapping of Venus and Mars
1:15,000,000	Conformal	Synoptic mapping of all planets
1:10,000,000	Equal-Area	Synoptic mapping
	Conformal	Synoptic mapping
1:5,000,000	Conformal	Systematic reconnaissance mapping; synoptic mapping of small bodies
1:2,000,000	Conformal	Systematic mapping; synoptic mapping of small bodies
1:1,000,00	Conformal	Systematic mapping
1:500,000	Conformal	Systematic mapping of areas of special interest
>1:500,000	Conformal	Special-area mapping

The selection of the 1:5,000,000 scale was originally based on the resolution of Mariner 9 images of Mars. Planetary geologists had found that five to eight pixels are required to identify and classify most landforms. Halftone screens used to print maps do not allow clear representation of features that have dimensions much smaller than 0.5 mm. The nominal resolution of more than 20 percent of the Mariner 9 images of Mars is 0.5 km per pixel. Features having minimum dimensions of 2.5 km could therefore be classified geologically, and reproduced at ten pixels per mm, or 1:5,000,000 (Batson, 1973). This approach has been successful; several map series have been based on it, and geologists and other map users have been able to match interpretive details to maps and spacecraft images alike. The resolution of images taken during other early reconnaissance exploration of most planets has usually been similar to that of Mariner 9 images. Thus, the scale of 1:5,000,000 has been selected for systematic reconnaissance mapping of all planets. As map scale becomes larger, the number of quadrangles required to cover a planet increases exponentially. For example, all of Mars has been mapped on 140 quadrangles at a scale of 1:2,000,000; a fourfold increase in map scale (1:500,000) would require approximately two thousand sheets. Although many studies require this detail, and Viking Orbiter images could support such mapping, fiscal considerations limit map compilation to selected maps in this series.

Consistent scales within map series are necessary for comparison of features. For example, mapping part of a planet at a scale of 1:5,000,000, part at 1:3,500,000, and part at 1:2,135,000 does not permit ready correlation of regional work into a global context. Consequently, scales are selected on the basis of the highest-resolution images available for most of a planet, which results in the enlargement of some image areas far beyond the ten pixel per mm optimum scale.

The number of different nominal scales is restricted to a minimum (Table 3.1). Thus, there is no 1:3,500,000 series; rather, a scale of either 1:5,000,000 or 1:2,000,000 is used, based on image resolution and map requirements.

No specific rules have been developed for maps at scales larger than 1:500,000 because so few planetary data are available to support these larger scales.

A variety of maps, particularly of the Moon and Mars, were produced prior to the development of consistent, systematic formats. For example, a series of 1:5,000,000-scale maps were made for multiring basins on Mercury, the Moon, and Mars. The Oblique Stereographic projection was used for these maps. Other early maps include special-purpose high-resolution maps on Mars, most of which could now be accommodated by either the 1:2,000,000 or the 1:500,000 Mars Transverse Mercator (MTM) system or by planetwide map formats.

3.2.   THE MAPPING PROCESS: PAPER MAPS

Making planimetric planetary maps after map controls are derived (see Chapter 6) involves digital processing of images transmitted by spacecraft, assembling the images into photomosaics, and in some cases manually painting a picture of the planetary surface on a map projection with a tiny spraygun, or airbrush. No single version of a map can serve all scientific investigations. Some versions emphasize surface relief, others emphasize surface markings or albedo. Several versions of many kinds of maps are therefore typically produced.

3.2.1.   *Image processing*

The techniques of image processing are performed with computers and computer-driven film-writing devices and involve the correction of digital images by changing the density number (DN) of each image element according to some defined mathematical model. A DN is a number (usually between 0 and 255, with eight-bit encoding), that specifies a pixel's level of gray. Image-processing methods were originally developed by space scientists at the Jet Propulsion Laboratory to enhance digitized lunar images returned by the Ranger spacecraft, and most subsequent development has been in support of planetary missions. The techniques are, however, used extensively by medical technicians to enhance x-ray photographs, remote sensing scientists to enhance images of the Earth, and by other specialists.

Preliminary image processing is performed on signals as they are received from spacecraft. This is a decoding process in which image data are separated from other signals, lines of data are identified and placed in their proper order, and the image is transmitted to an image display tube and a film writer.

Systematic "batch" processing produces two or three versions of each frame. This processing is done automatically, according to predicted models of space-craft and camera behavior, and with little human intervention. The first version (Figure 3.6A) is corrected for camera shading, contrast is enhanced, and reseau marks are removed. Radiometric corrections are made according to prelaunch calibrations. High-frequency image detail is enhanced by spatial filtering on

A

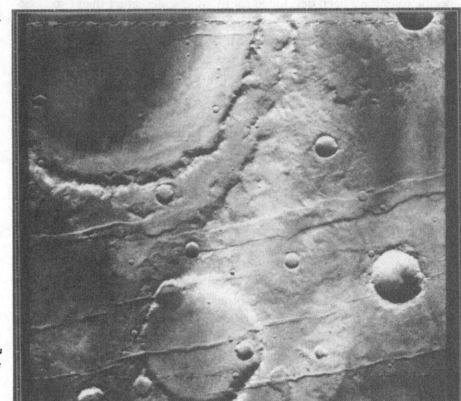

Figure 3.6
A. Shading-corrected image of Mars. The shading and reseau patterns have been removed in the computer. The large crater in the upper left corner of the frame is called Slipher (Viking Orbiter PICNO 535A22 SCR 2 rectilinear; JPL).

a second version (Figure 3.6B). A third version (Figure 3.6C) is geometrically corrected for camera distortion and rectified to an orthographic (overhead) view of the planet according to the predicted orientation of the camera with respect to the surface. This correction is usually performed on the spatially filtered version.

3.2.2. Photomosaics

Many kinds of preliminary and interim photomosaics are produced. They differ from true maps because photographic images are collections of complicated light and dark patterns related to both illumination and surface coloration and thus do not lend themselves to precise verbal or schematic definitions. Geometrically, they are perspective views, and distortions are contributed by the optical and electronic systems used to record them. They are projections of three-dimensional objects and surfaces onto two-dimensional image planes. Mountainous terrain, therefore, is presented to the viewer as a complicated mixture of planimetric and profile views.

Four basic types of photomosaics, defined according to their geometric characteristics, are used in planetary science. Except for unusual circumstances, only controlled photomosaics and orthophotomosaics are published in the

B

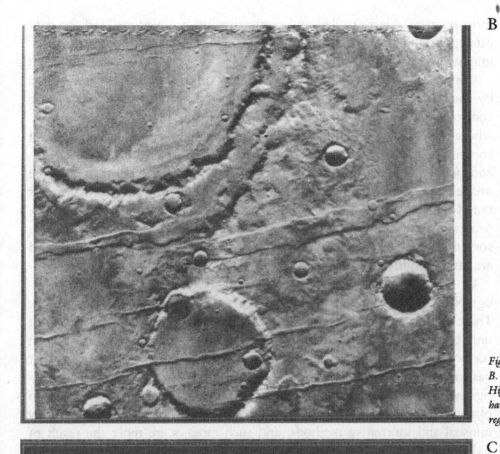

Figure 3.6
B. A spatially filtered image.
High-frequency image details
have been emphasized, but broad
regional variations are subdued.

C

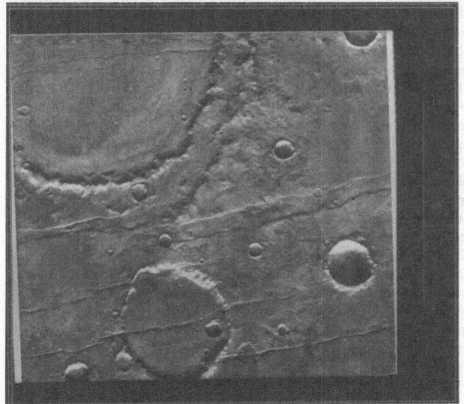

Figure 3.6
C. Orthographicaly projected im-
age used in making maps.

form of maps. Uncontrolled and semicontrolled mosaics are used for data cataloging and interim scientific reporting, but they are not formal cartographic products.

Controlled photomosaics are made after a control net has been compiled by analytical photogrammetry (see Chapters 5 and 6). A photogrammetric control net consists of a set of well-defined image points (usually craters) whose latitudes and longitudes have been computed precisely. The process of computing this control net results in revised values for camera orientations and positions, which can be used to improve geometric correction of the image frames so that mosaics can be made that are not only internally consistent but also match the computed positions of map control points. This complex process typically requires years to complete. Controlled photomosaics are sometimes used as base maps for the compilation of more refined maps and are often published as formal map products.

Orthophotomosaics of the Earth are made routinely because detailed topographic information can be derived during the map compilation process. These highly complex and specialized mosaics consist of images that have not only been placed precisely with respect to map controls, but have also been corrected for the distortions inherent in any perspective view of a surface that has topographic relief. Routine compilation of orthophotomosaics requires rigidly controlled image-gathering conditions that are very rarely possible in planetary exploration. Planetary orthophotomosaics can only be compiled by unusual and innovative methods, and such mosaics are more the result of individual research projects than of routine map production processes. Although a series of 1:250,000 Lunar Topographic Orthophotomaps (LTOs; Table A-I.10) of a small part of the moon were made, few other orthophotomosaics have been made for extraterrestrial surfaces.

Uncontrolled mosaics (Figure 3.7) are made as soon as possible after images are received from a spacecraft and photographically reproduced – usually in a matter of hours. The images will have been digitally processed to the extent required to produce reasonably clear photographs but will contain geometric and radiometric distortions. Each image is a perspective view. Those taken at oblique angles to the surface of the planet are foreshortened. Frames in uncontrolled mosaics are joined by matching overlapping images but are unlikely to match very well, resulting in mosaics that have large errors and discontinuities. No attempt is made to match photographic tone or contrast in adjacent prints.

Semicontrolled mosaics are made from orthographically rectified, spatially filtered images produced during the systematic image-processing phase. Photographic prints are made at scales predicted by tracking data, which give the expected location and orientation of spacecraft and camera. They are placed in mosaics according to tracking data predictions and modified for a best fit to details appearing in adjacent, overlapping images (Figure 3.8A, B).

The geometric accuracy of semicontrolled mosaics is affected by several factors. Orthographically projected images are not quite the shape they would

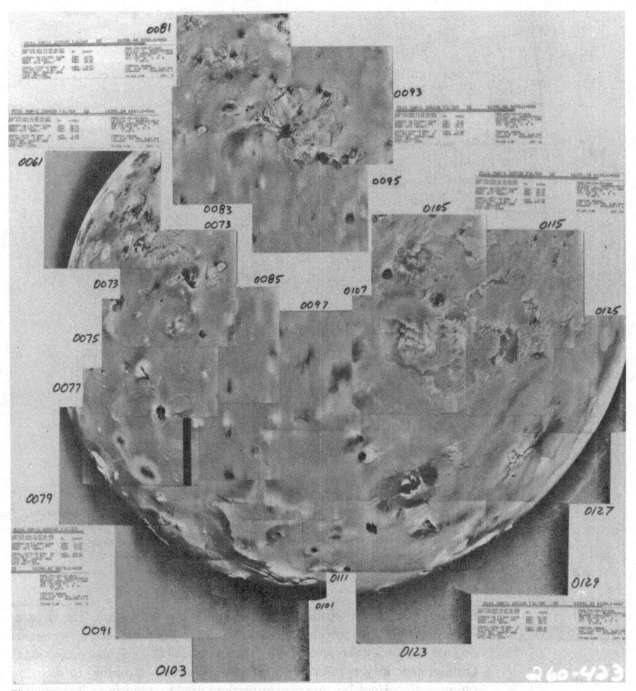

Figure 3.7
Uncontrolled mosaic of Io images taken during the Voyager 1 mission. Significant mismatches of images result from the rapid scale changes as the spacecraft approached Io. Some of the frames are blurred because the high speed of the spacecraft could not be adequately compensated (mosaic 260-423; JPL/USGS).

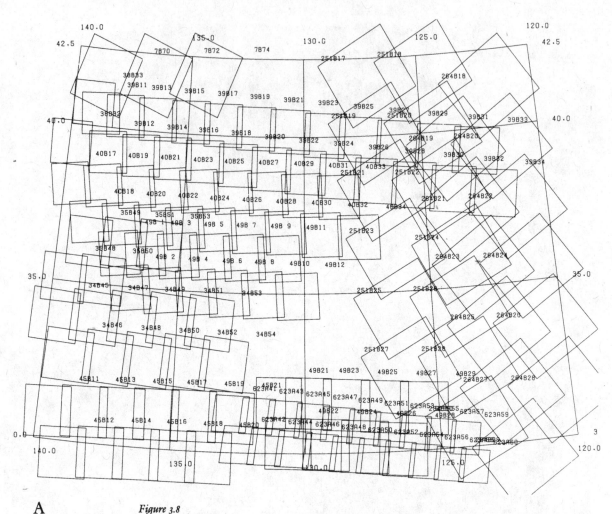

A

Figure 3.8

A. Computer-drawn "footprint plot" made on the basis of spacecraft trajectory information, the astronomical ephemeris of the target planet, and spacecraft scan-platform pointing information. Although the relative positions of frames within each orbital revolution are quite accurate, slight errors in knowledge of spacecraft and camera orientation result in absolute errors as large as one-third the width of a frame.

be if they had been transformed to the specific map projection. The error is significant in images covering very large areas on the planet but is negligible on images covering very small areas if they are placed in conformal projections. Orthographic images must be scaled in the photolab according to their location on the planet, because map projections used for semicontrolled mosaics vary with latitude, whereas the orthographics have true scale in the center of the frame.

No attempt is made on semicontrolled mosaics to conceal the boundaries between frames, either by "feathering" frame edges or by selecting cutlines to follow natural lineations. The intent is not only to produce the mosaics quickly, but to avoid introducing artifacts that might be mistaken for natural surface features.

Figure 3.8

B. Semicontrolled photomosaic of the southeastern part of the Diacria (MC-2) quadrangle of Mars. The mosaic was made from Viking 1 images of Mars and was controlled by reference to an earlier, lower-resolution map of the area made from Mariner 9 images, the footprint plot in 3.8A. This is an intermediate map used for image-data cataloging and preliminary scientific analyses.

B

Controlled mosaics are tied to derived map controls, rather than to image locations predicted by spacecraft tracking. The manual compilation of controlled mosaics is identical to that of semicontrolled mosaics, except for the method of controlling the placement of individual frames. Uncontrolled, semicontrolled, and many controlled mosaics are made manually, simply by sticking paper prints together on some kind of base material. However, more controlled mosaics are being made by assembling digital images into arrays in a computer.

Manual mosaic compilation is an iterative process in which picture scale is assumed, pictures are printed to scale, and a mosaic is made. The inevitable errors are evaluated, a new scale is selected, and pictures are reprinted and remosaicked. Three or four reprintings of large numbers of frames is common

Figure 3.9
Patricia Bridges checking tones and densities in an airbrush map with a densitometer.

practice in controlled mosaicking (much to the annoyance of photolab technicians!), whereas the first iteration is usually accepted for semicontrolled mosaics.

3.2.3. *Airbrush maps*

Cartographers have long experimented with ways to show landforms on paper but have rarely achieved a satisfying and natural representation. Photomosaics rarely serve the purpose because of inconsistent illumination, variable resolution, and obscuration of relief by albedo markings. Only a single version of a single image can be used in any given photomosaic, although many pictures showing a diversity of information may exist. Consequently, much available information is often excluded from photomosaics. Topographic contour lines, where they can be drawn with sufficient resolution and detail, show an experienced map reader the three-dimensional geometry of the land distinctly, but available data are rarely adequate to draw sufficiently detailed contour maps of planetary surfaces.

Shaded relief ("hill-shaded") maps present the clearest pictures of surface morphology in the absence of high-resolution contour maps, but their compilation cannot be automated. It must be done by specially trained cartographers skilled in the use of an airbrush (see Chapter 2, Figure 2.10). This method can produce a composite painting that shows essential information contained in large and awkward datasets (Inge, 1972; Inge and Bridges, 1976; Batson, 1978). The airbrushing technique used in planetary cartography requires artistic talent, skill in the use of the airbrush, and most importantly, a talent for visualizing landforms by examining many different photographs with widely varying scale, surface resolution, viewing angle, and obscuration by clouds or haze. After a mental image based on this examination has evolved, the airbrush cartographer must be able to draw that image so that it is clearly visible to others. The technique is invoked only for specifically defined tasks, not because it is expensive, but because so few trained practitioners are available that their labors must be judiciously rationed.

Airbrushing was originally devised to help extract information from sparse data. Prior to lunar spaceflights, modern airbrush maps of the Moon were based on mosaics of photographs taken through telescopes. Cartographers made sketches based on these mosaics and refined them by viewing the Moon through telescopes. Not only were they able to see more detail than could be photographed, but they were able to watch the surface as its illumination changed, observing features that appeared only for an instant under the best seeing conditions and illumination. They were thereby able to make portrayals impossible to achieve from photographs alone (Kopal and Carder, 1974).

This approach had similar application to the mapping of Mars from Mariner 9 data during the early 1970s and to the more recent mapping of outer planet satellites from Voyager data. Although visual telescopic observation was not feasible, the meticulous examination of a variety of images resulted in maps displaying a level of detail that could not have been produced in any other way.

The first step in the making of an airbrush map is the "lay-in." After gleaning all usable details from the mosaic, the cartographer places a white backing sheet between it and the manuscript and examines other data to add detail to the drawing. In the final step an electric eraser is used as a drawing instrument to modify tones and accentuate highlights.

Uniformity and consistency are essential in portraying landforms with the airbrush. These are maintained by intensive review of each drawing, first by cartographers trained in airbrushing, then by scientists familiar with the area. Final map revision consists of responding to reviewers' comments and adjusting map tones to publication standards by checking sample areas with a reflection densitometer (Figure 3.9). Highlights, average midtones, and dark areas are adjusted for compatibility with printing processes (Appendix II).

Preliminary airbrush maps are made in conjunction with uncontrolled mosaics during spacecraft mission operations. They are made without benefit of a controlled photomosaic base because such a base is almost never available at this stage of mapping. Instead, perspective grids based on spacecraft trajectory and planetary ephemeris data are drawn for image frames used to control the map. The cartographer then sketches a map on the proper projection, transferring image details from the perspective grid to a map graticule, grid cell by grid cell. Once this process is complete, details are added, with more

Figure 3.10
Small-scale mosaic of high-resolution images of the Syrtis Major region of Mars.

characterization and generalization of forms than would be permitted in a final map.

These maps must be made within a few weeks of data reception to support preliminary scientific analysis and reports by mission science team members. One of the first of these preliminary maps was made from Mariner 9 images taken of the south polar region of Mars. A global dust storm had enveloped the planet when the spacecraft arrived, and for more than a month the storm frustrated the efforts of scientists to gather useful pictures of the surface. As patches of relative clearing moved across the south polar region, it was possible to use bits and pieces of several pictures to assemble a coherent view of most of the area. Preliminary, uncontrolled airbrush maps have been made for all planetary missions since that time. Each of these maps requires an average of one hundred hours of compilation time.

Synoptic and systematic maps of planets are made with the airbrush because photomosaics are difficult to interpret on a regional scale. Low-resolution, small-scale images show regional coloration, but they do not show relief information, because the slopes that delineate landforms are not resolved. High-resolution images, on the other hand, emphasize topographic details equally, be they regional or local. The Voyager images of the outer planet satellites, for example, are collections of high-, low-, and intermediate-resolution pictures. Regardless of processing, mosaics of these images do not make very good maps. Similarly, mosaics of large numbers of high-resolution images (Figure 3.10) are poor because surface details are not uniform and because artifacts such as cutlines produce a "fish-scale" effect that dominates the map. The problems can be resolved with airbrush maps (Figures 3.11 and 3.12).

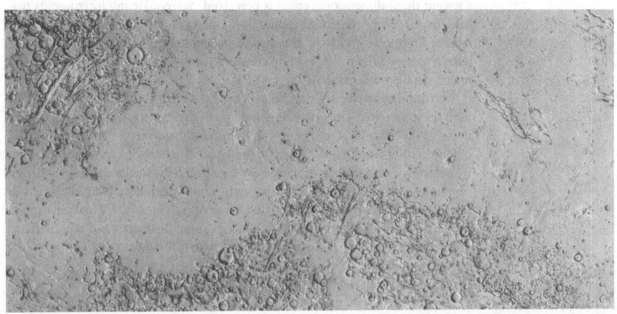

3.2.4. Topographic maps

Topographic contour maps of the Earth are made from stereoscopic aerial photographs by a process called photogrammetry (see Chapter 5). The photographs are taken with specialized, virtually distortion-free, wide-angle film cameras from aircraft. Stereoscopic photographs are made by taking the pictures at uniform intervals while the aircraft flies at a constant speed. The exposure intervals are selected so that each picture overlaps the previous one by about 60 percent. When the overlapping images are viewed stereoscopically,

Figure 3.11
Airbrushed, shaded relief map of the Syrtis Major region of Mars. This map shows only landforms; albedo information has been excluded.

Figure 3.12
Airbrushed map showing albedo information superposed on shaded relief.

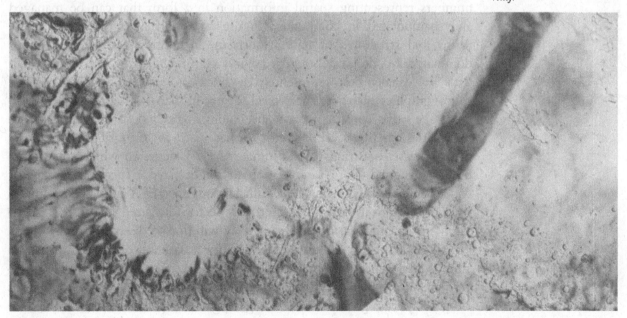

a strong three-dimensional effect is perceived. Stereoplotting instruments utilize this effect for precision mapping by coupling an index mark with a plotting device. An operator viewing the three-dimensional "model" on a stereoplotter is able to move the index mark in measured amounts, allowing the tracing of image details onto maps or digital files.

Vertical measurement accurate to 1/10,000 of the flight altitude is possible with these systems. Thus, a stereoscopic pair of pictures taken from an aircraft flying at an altitude of 10,000 m can be used to measure the relative altitudes of objects within ±1 m. Although much of this technology is available in planetary science, the cameras used for aerial surveys usually cannot be carried on planetary spacecraft. The only exception was the metric camera carried in orbit around the Moon during the last three Apollo missions. This film camera was a half-scale version of standard aerial survey cameras and was used to make highly detailed contour maps of a part of the equatorial region of the Moon.

3.3. THE MAPPING PROCESS: DIGITAL MAPS

The word *map* calls to mind a piece of paper with various symbols and notations indicating features. The appearance of a map is designed with the dynamic range of the human eye in mind. This format restricts mapping with most forms of remote-sensing data, which are gathered in wavelengths, dynamic ranges, and intensity levels undetectable by the human senses. For example, an eight-bit digital television image discriminates 256 shades of gray and commonly contains far more information than can be discriminated by the eye, by the television screen, or by film (which are limited to twelve to sixteen shades of gray). Digital maps consist of files of numbers representing spatial information in a form that can be managed and manipulated in computers (Figure 3.13). Planetary maps depict the forms and structures of natural surfaces, in contrast to terrestrial maps dominated by political boundaries and cultural features. Terrestrial map data are usually best represented by lines and symbols, whereas image (photographic) maps are the typical format in planetary cartography. Digital image maps consist of rows and columns (rasters) of brightness values (intensities) or pixels. The "vector format" used for many digital terrestrial maps consists of strings of coordinate (X, Y) values to which some attribute is assigned. For example, a road might have a specified designator code, followed by a string of a sufficient number of X and Y coordinates to depict the route taken by the road. Vector-formatted maps are used so rarely in planetary cartography that they will not be discussed here.

Although there will always be a need for a basic set of printed maps, the time is near when the total number of map versions required will constitute an intractable collection if printed in the traditional way. Many kinds of geophysical, geochemical, and geological data can be transformed to

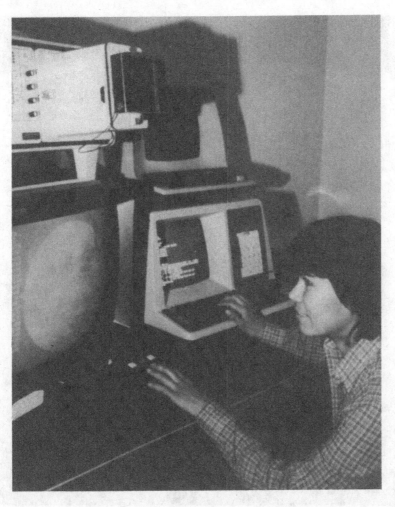

Figure 3.13
Ella Mae Lee begins compilation of a digital image map on an interactive image-processing terminal.

map or image format and can profitably be registered with digital image mosaics or other data arrays, further increasing map complexity (Figures 3.14 and 3.15). Not only has the number of required map combinations increased far beyond the level at which it is feasible to use traditional publication methods, but many of the versions may be important to only a few specialists. Furthermore, some users of maps may be unaware of the scientific potential of certain combinations of maps unless they actually attempt to use them.

There are important differences between digital image maps designed for efficient computer storage and access and photographic image maps used for viewing and analysis. Only the numbers representing pixel locations and intensities – and the efficiency with which a computer can access those numbers – are relevant for images stored as pixels in a computer file. For the eye and brain to make sense of them, however, the files must be reconstituted as two-dimensional pictures. These may take the form of small images or parts of digital maps displayed by the computer on television screens or maps and

Figure 3.14
The multidimensional digital image map. A collection of digital maps in the same format can be combined in various ways to demonstrate the spatial relationships of various aspects of the surface of a planet. In this illustration the top layer is a small-scale digital geologic map of Mars; the next layer is a color-coded map of topographic elevations on that planet, and the third is a digitized image map. A multitude of other digital maps makes up the remainder of the "stack."

"hard-copy" images transmitted by the computer to special film-writing devices and reproduced photographically.

Digital maps are more versatile than printed maps because the data can be compiled without being tailored to specific projections and scales, and one can make many kinds of customized conventional maps from the same basic datasets. Computers can be used to extract from a digital array data lying within any specified map area, to transform them to any desired map projection, and to magnify or compress them to any desired scale. The enhancements and combinations that most clearly illustrate the results of an investigation can be selected by interactive experimentation on image-display devices prior to publication. These customized maps can then be written on film for publication as conventional maps.

Figure 3.15
Composite digital-image map showing the geology of the western hemisphere of Mars (Scott and Tanaka, 1986) superposed on a shaded relief map base.

Until recently, the greatest barrier to digital cartography has been computer storage capacity. A mosaic can easily contain 10,000 lines by 10,000 samples, or 100 million image elements. The compilation of such a mosaic requires several times this storage capacity to allow interim storage of working arrays. As mass storage devices and computer memory become more available, the compilation and manipulation of digital maps will become available to most investigators. Although digital compilation of planetary maps is now becoming routine, detailed specifications and compilation procedures for digital maps are being verified and modified even while the first systematic digital maps are being produced (Batson, 1987).

3.3.1. Digital planetary map formats

Digital planetary maps are designed to show the intensity of some specified attribute (reflected light or topographic elevations, for example) at each point on the surface of a planet. Digital television images returned by planetary spacecraft are first approximations of image maps. These are perspective views, however, containing a variety of spatial distortions and instrument errors. They can be made into true maps only by complex processing.

Digital terrain models are arrays of elevation, rather than brightness, values. They are usually made by photogrammetric measurement of stereoscopic images (see Chapter 5).

The fundamental process in making maps is the determination of correct locations of data points that delineate features. This requires geometric modification of each image, including correction of camera distortions and projection of each image to its correct ground location. If the goal is to produce traditional photomosaics and shaded relief maps, digital processing is complete with the solution of these geometric problems. Digital images, however, contain information regarding the light-reflecting properties of the surface that is important to planetary scientists. This kind of information can only be conveyed with digital computer files, not with lithographic images printed on paper. Exploiting these data requires correction of instrument and transmission errors so that the intensity of each pixel on an image map is an accurate indication of the radiance received by the detector. These corrections also have an important cosmetic effect on photomosaics. Properly made, the seams are subdued, giving the impression of a single image rather than a discontinuous montage. The compilation of digital planetary maps therefore includes processing that not only results in cosmetically acceptable photomaps, but processing that provides researchers with the most error-free array of image data that can be made from current technology.

Projection and scale are fundamental design considerations for all maps. The Sinusoidal Equal-Area map projection is used for digital planetary maps because it is easily manipulated both as a computer file and as a human-readable image. Each image line is a parallel of latitude, the length of which is compressed by the cosine of that latitude. The entire database for a planet can be contained in this single array, so projection boundaries need not be considered during processing. Resolution is uniform throughout; apparent distortion near the edges of the global images can be removed simply by sliding image lines with respect to one another until the meridian passing through the area of interest becomes a vertical line through the image (Figure 3.16). Any other transformation requires geometric resampling, which is not only more expensive, but also results in loss of image resolution.

The "scale" of a digital map is specified in terms of pixel size. This is not equivalent to "map scale," which depends upon the size of a printed reproduction of the digital map. Pixel dimensions are given in terms of solid angles from the center of a planet, rather than in terms of meters or kilometers per pixel. Because each image line is a parallel of latitude on the Sinusoidal projection, only rational fractions of degrees are considered when selecting scales, so that the distance from pole to pole is an integral number of lines of data.

Figure 3.16
Images of a low-resolution digital image map of Ganymede, on a Sinusoidal Equal-Area projection. Pixel dimensions are 1/16° by 1/16° at the equator, or approximately 2.9 km × 2.9 km. The central meridian is at longitude 0° in the first view, but the image is modified by sliding image lines with respect to each other in subsequent views to minimize viewing distortions in different parts of the image.

Changing the scale, or "resolution,"* of digital maps is easily accomplished by compression or magnification by powers of two; that is, by successively doubling or halving the scale. A digital map may thus have a scale of 1/8°,

---

* The word "resolution" is often used imprecisely as a synonym for *pixel size*. Resolution is more properly expressed in terms of modulation transfer function, a complex concept that is not particularly relevant to our discussion.

1/16°, or 1/32° per pixel for very low resolution maps, through 1/1,024° or 1/2,048° per pixel for very high resolution maps. If a 1/2,048° map is to be registered with a 1/1,024° map, the 1/2,048° map is simply compressed by a factor of two. A global map having a scale of 1/32° per pixel will contain 32 × 180 = 5,760 lines of data, and the equator will be 32 × 360 = 11,520 pixels (samples) long. This integral compression and expansion is rapid and economical, and does not involve geometric resampling.

The planetary surface area covered by each pixel in a database is constant. The longitudinal extent of any pixel is expanded by the inverse of the cosine of the latitude to maintain this condition. For example, a pixel at latitude 60° N might have dimensions of 1/64° latitude by 1/64°/cos 60 = 1/32° longitude, whereas at the equator the dimensions would be 1/64° latitude by 1/64° longitude. These pixels will have the same linear dimensions in either case because meridians converge toward the pole. For example, on Mars, 1/64° of longitude or latitude is approximately 925 m at the equator, but 1/64° of longitude is only 462 m at latitude 60° N.

3.3.2. *Compilation phases*

Compilation processing is done in four stages. The digital file resulting from each stage is preserved as a basis for further processing and for specialized analytical use. These stages are defined as follows.

LEVEL 1   This stage of processing (Figure 3.17A) produces a radiometrically accurate image free of artifacts and noise and at the best resolution possible. No geometric correction is done, because geometric correction degrades resolution. Images that have been processed through level 1 make the best pictures for interpretation of surface details.

Level 1 processing removes non-geometric errors contributed by the camera optical, electronic, and transmission systems. These include radiometric instrument errors, camera shading, interference patterns caused by other spacecraft instruments, and data errors and dropouts introduced during transmission to Earth. Data obliterated by reseau marks or lost during transmission cannot be restored, but their disruptive effects on images can be reduced by interpolating new values for "lost" pixels from surrounding image data.

Camera shading and gain are akin to base fog on film; they can be detected by taking pictures periodically during a mission of black space (see Chapter 1, Figure 1.5) and of bright, featureless areas such as dust clouds on Mars, to produce correction images that can be "subtracted" from data images. Shading varies with time, with the age of electronic components, with the electronic and optical peculiarities of each camera, and with high levels of radiation such as those encountered in the vicinity of Jupiter. Interference patterns, dust specks, and data errors can be mapped in most low-contrast images and removed by digital processing. All processing information, including the locations of reseau marks, is recorded and stored in the digital label of each image to support subsequent processing levels. Special versions of selected level 1 images, in which the reseau pattern is retained, are made as required for use with stereomapping equipment.

LEVEL 2   Geometric corrections are made at this stage of processing (Figures 3.17B, 3.18). They involve making a raster in which the location of each pixel is defined by latitude and longitude. Processing at this level requires that camera distortions be corrected and that each image be projected to an array wherein the line and sample coordinates of each pixel in the raster are latitude–longitude coordinates. Camera distortions are corrected by reference to the reseau pattern. The pattern is visible on images returned by the spacecraft and can be used to correct transmission distortions by moving reseau images so that they correspond with their calibrated locations. This is done with a second-order polynomial model that produces results that are accurate to less than one pixel for all but the extreme corners of the image.

Projecting each image to its correct ground location requires precise knowledge of (1) the location of the spacecraft, (2) the location of the planet, (3) the orientation of the planet, and (4) the pointing orientation of the spacecraft camera. The location of the spacecraft is derived by tracking its radio signals. The motions and locations of planets and satellites are derived by a combination of spacecraft radio-tracking and astronomical observations and are known quite accurately. The pointing angles of spacecraft cameras are measured by star sensors and other devices on spacecraft and transmitted to Earth when pictures are taken. These factors can be used to compute approximate locations of pictures, but they are not accurate enough for direct cartographic compilation because each source of information is subject to error. The preliminary locations are therefore refined by photogrammetric triangulation with overlapping images on adjacent frames (Chapters 5 and 6). Accuracy is improved dramatically if overlapping frames of the entire planet are available. Triangulation computations result in accurate camera orientations for those frames used in the computation and in a control net of coordinates for selected features on the planet. Frames used for mapping, however, frequently are not those used for the triangulations. A new set of camera orientations for the mapping frames must therefore be derived during map compilation (Edwards, 1987). This final computation assumes that orbital locations of camera stations are accurately known, and it modifies camera orientations until images correspond with previously derived control points and with adjacent frames in the map. The level 1 images are projected to their predicted locations on the surface of a planet on the basis of the newly derived orientations. Revised camera-pointing angles are recorded for future use and for possible reevaluation of ground control networks. The pixel (digital scale) size of level 1 images is preserved in the level 2 stage.

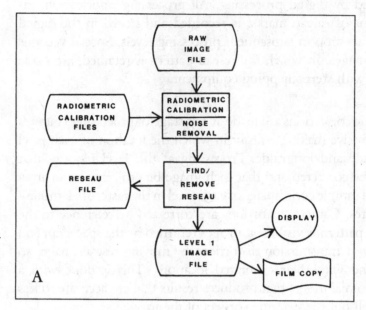

Figure 3.17
A. Level 1 processing flow. Radiometric calibration files are derived by analysis of images of sky and of diffuse surfaces taken in flight.

LEVEL 3   Although reasonably precise radiometric and geometric corrections can be made for spacecraft images, accurate correction of photometric and haze effects is far more elusive. Surface materials reflect light differently, depending upon mineral composition, texture, illumination, etc. Join lines between adjacent pictures in a photomosaic are typically visible and distracting even if perfect radiometric and geometric image corrections can be made (Figure 3.19). Visible cutlines in a mosaic detract from cosmetic appearance and interpretability. They produce a kind of screen, or "fish-scale" appearance, in a mosaic that tends to obscure subtle regional surface phenomena that might otherwise be distinct. Perhaps even more significantly, visible cutlines indicate incomplete or incorrect radiometric or photometric processing of individual images. Level 3 processing is intended to compensate as much as is practical for these effects. Should the photometric model be revised, reprocessing at levels 1 and 2 is not required.

Various reflection laws (photometric functions) have been derived for extraterrestrial surfaces. The Minnaert (1941) function

$$B = B_o(\cos I)^{k(a)}(\cos E)^{k(a)-1} \tag{1}$$

has been used with good results to correct tones and contrasts in images within mosaics. In this equation, B is the reflectance of a surface of normal albedo B_o, viewed at an angle E from normal, illuminated at a solar zenith

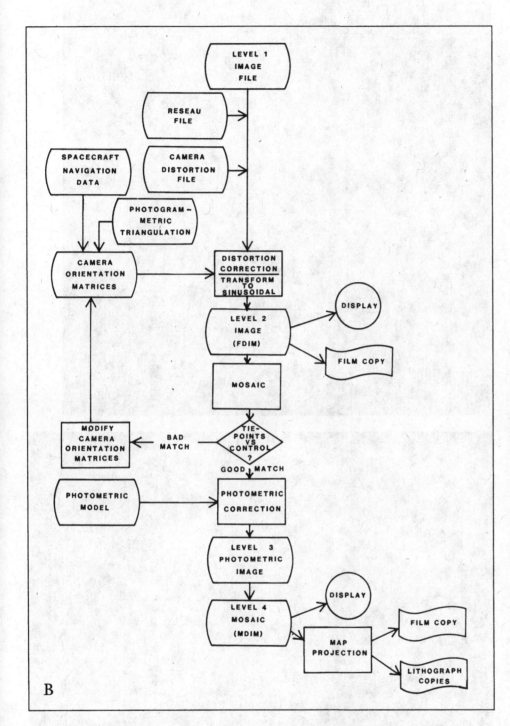

B

Figure 3.17
B. Processing flow for levels 2, 3,
and 4. Reseau files contain the
image coordinates of reseau
marks determined during level 1
processing, as well as their true
positions, determined by preflight
measurement. Camera distortions
are measured prior to launch
and then during flight by measur-
ing star positions. Optical distor-
tions are negligible, but electronic
distortions are significant. Initial
values in camera-orientation
matrices are based on spacecraft
radio tracking and optical navi-
gation and on photogrammetric
triangulation. They are verified
or modified during mosaicking on
the basis of how well images fit
together. Photometric models are
based on Minnaert reflectance
rules and modified by analysis of
tonal discrepancies between over-
lapping images.

angle I, and $k(a)$ is a coefficient that varies with phase angle a. The value of k is modified as required by empirical evidence. Given a choice, cartographers select frames taken from the same orbital revolution for a photomosaic because photometric effects are similar in adjacent frames, simplifying cosmetic cor-

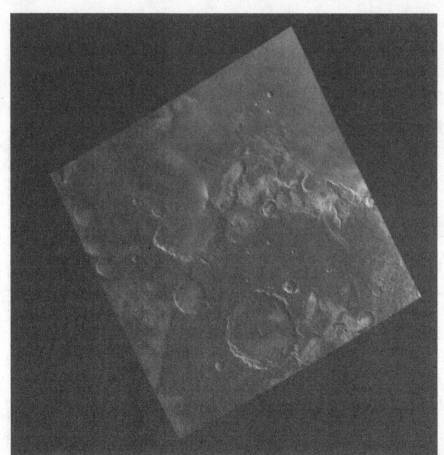

A

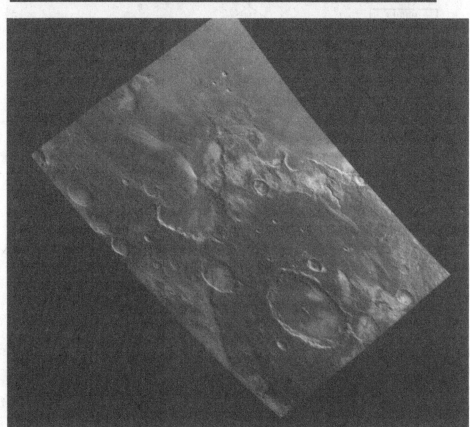

B

Figure 3.18
A. Level 2 image of part of the
Amazonis Planitia of Mars in
Sinusoidal projection; local cen-
tral meridian at longitude 159°
W. This image has been modified
for viewing by shifting the cen-
tral meridian (Viking Orbiter
Picno 637A75).

Figure 3.18
B. The same image with a central meridian of 0°.

Figure 3.19
Digital (Sinusoidal) mosaic of level 2 Voyager images of Rhea, a satellite of Saturn. Photometric corrections have not been made, resulting in tonal mismatch between adjacent images.

Figure 3.20
The mosaic of Figure 3.19, but with photometric (level 3) corrections in each frame.

rections. A discussion of radiometric and photometric correction of planetary images is given by Soderblom and others (1978) and by McEwen (1988).

When a cosmetically attractive product is required for publication, the individual images in the mosaic can be "high-pass filtered." This is an image-processing technique that produces uniform image tones for mosaicking but destroys the radiometric integrity of the image. These filtered mosaics are typically made for publication, whereas accurate photometric correction is pursued for digital map archives.

LEVEL 4   This is the final image mosaic (Figure 3.20). Determination of the optimum placement and overlap of images in a mosaic is a complex iterative process, and can be done with level 2 images, before level 3 processing is complete. After mosaicking parameters have been derived and after level 3 processing is complete, the level 3 images are substituted for the level 2 images in the mosaic. Whereas level 2 images are processed at the highest possible resolution, all frames in the level 4 mosaic are normalized to a single value by compressing highest resolution frames to the "average" for the mosaic. For example, most level 2 images in a level 4 mosaic might have pixels with dimensions of $1/256°$ and a few of $1/512°$. Each 2×2 square of pixels in the latter images would be averaged to produce $1/256°$ pixels, for uniformity with the rest of the mosaic.

3.4. MAPPING THE SMALL, IRREGULARLY SHAPED SATELLITES AND ASTEROIDS

Maps of spherical or spheroidal planets are familiar and well understood. As projections of curved surfaces on sheets of paper, they have mathematical properties that facilitate the measurement of areas or distances. Many of the small satellites and asteroids of our solar system, however, are irregular in shape. Conformal and equal-area map projections of such objects are difficult not only to define and compile, but also to understand and use. Analytical work involving volumes, dimensions, and spatial distributions of materials on these bodies is best done with digital maps.

A digital map of an irregular object consists of a Sinusoidal Equal-Area array of radius measurements registered with an array of images transmitted by spacecraft. It is made primarily by photogrammetric computation of radii at conspicuous features on the object (Duxbury, 1974; Duxbury and Callahan, 1989). Additional radius values between these control points can sometimes be added with stereophotogrammetry. Spacecraft data-gathering sequences and flight trajectories rarely allow systematic photogrammetric surveys, however; most datasets consist of collections of images, which have large differences in illumination between images in otherwise acceptable stereopairs, and pairs with such large convergence angles that stereoscopic viewing is impossible. In most cases, therefore, the radius grid is simply filled by interpolation.

It is also possible to sculpture a model of a satellite or asteroid by projecting a spacecraft image onto an approximate model. The model is compared to the image, and sculptured as necessary. The process is repeated with images taken from different perspectives, until the visual appearance of the model matches that of the images. The analog process, demonstrated by Turner (1978), results in an impressive but cumbersome solid model. The corresponding digital process results in a collection of latitudes, longitudes, and radii in register with each spacecraft image used in the compilation (Figures 3.21, 3.22, and 3.23).

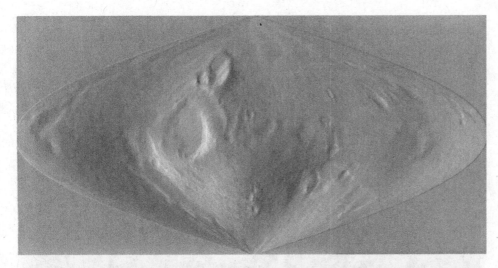

Figure 3.21
Digital shaded relief image of a Sinusoidal Equal-Area projection of a radius model of Phobos.

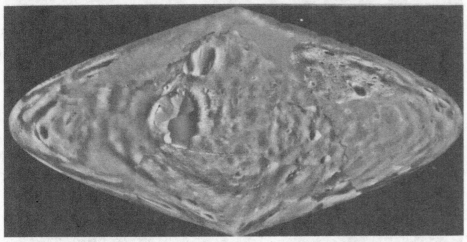

Figure 3.22
Digital image mosaic of Phobos, on the Sinusoidal Equal-Area projection.

3.5. SCOPE OF DIGITAL PLANETARY CARTOGRAPHY

Twenty planets and satellites have been explored by spacecraft at the time of writing. Although some analog maps have been made for all of these bodies, new and revised mapping by digital methods is under way (see Appendix III).

Satellites of the outer planets (Jupiter, Saturn, Uranus, and Neptune) were mapped from only a few hundred Voyager images, whereas the two Viking Orbiters returned approximately 55,000 high-resolution images of Mars. Nearly all the images of the satellites of the outer planets will be used for digital mapping, but only a subset of the Viking image dataset will be used for Mars.

The total storehouse of spatial data returned by planetary spacecraft is so enormous that traditional cartographic products are inadequate to support

Figure 3.23
Orthographic mosaic of the southern hemisphere of Phobos. This is a digital reprojection of a three-dimensional photomosaic of Viking Orbiter images.

research in planetary science. These products can be compiled and distributed in tractable digital formats to complement carefully selected conventional paper maps. Individual scientists and others with access to the minimal digital image-processing capability now becoming available on personal computers will then find new and ingenious solutions to problems that have hitherto defied analysis.

3.6.   SUMMARY

Planimetric maps of twenty planets and satellites have been compiled. These consist of photomosaics, shaded relief maps, and digital mosaics. They are required as base materials for geologic mapping and for planning future missions.

Planetary maps differ in form and content from terrestrial maps in that planetary maps consist almost entirely of landform images, whereas terrestrial maps often are dominated by political boundaries and cultural symbols. Moreover, planetary mapping is heavily dependent upon digital processing of images returned by spacecraft.

Digital planetary maps consist of arrays of brightness, elevation, or other values, in which image lines are parallels of latitude and sample designations are longitudes. The pixel values in the arrays typically have a broader dynamic range than can be discriminated by the eye, so digital maps contain more information than paper maps. Digital maps can be printed with film-writing devices for publication by conventional photographic or lithographic processes, or they can be used analytically in computers.

3.7. REFERENCES

Batson, R. M. 1973. Cartographic products from the Mariner 9 mission. *J. Geophys. Res.* 78 (20): 4424–35.

Batson, R. M. 1978. Planetary mapping with the airbrush. *Sky and Telescope* 55 (2): 109–12.

Batson, R. M. 1987. Digital cartography of the planets: new methods, its status, and its future. *Photogrammetric Engineering and Remote Sensing* 53 (9): 1211–18.

Edwards, Kathleen. 1987. Geometric processing of digital images of the planets. *Photogrammetric Engineering and Remote Sensing* 53 (9): 1219–22.

Duxbury, Thomas C. 1974. Phobos: Control network analysis. *Icarus* 23: 290–9.

Duxbury, Thomas C., and Callahan, John D. 1989. Phobos and Deimos control networks. *Icarus* 77: 275–86.

Inge, J. L. 1972. *Principles of Lunar Illustration*. Aeronautical Chart and Information Center Reference Publication 72–1.

Inge, J. L., and Bridges, P. M. 1976. Applied photointerpretation for airbrush cartography. *Photogrammetric Engineering* 42 (6): 749–60.

Kopal, Z., and Carder, R. W. 1974. *Mapping of the Moon*. Dordrecht: D. Reidel.

McEwen, A. L. 1989. Photometric functions for photoclinometry and other applications. *Icarus* (in press).

Minnaert, M. 1941. The reciprocity principle in lunar photometry. *Astrophysical Journal* 93: 403–10.

Scott, D. H., and Tanaka, K. L. 1986. *Geologic Map of the Western Equatorial Region of Mars*. U.S. Geological Survey Miscellaneous Investigations Series Map I–1802A.

Snyder, J. P. 1982. Map projections used by the U.S. Geological Survey. U.S. Geological Survey Bulletin 1532.

Snyder, J. P. 1987. Map projections used by the U.S. Geological Survey. U.S. Geological Survey Professional Paper 1395.

Soderblom, L. A., Edwards, K., Eliason, E. M., Sanchez, E. M., and Charette, M. P. 1978. Global color variations on the Martian surface. *Icarus* 34: 446–64.

Stooke, Philip J., and Keller, Peter C. 1989. Map projections for non-spherical worlds: the variable-radius map projections. *Cartographica* (in press).

Turner, Ralph J. 1978. A model of Phobos. *Icarus* 33: 116–40.

4

Planetary nomenclature

MARY E. STROBELL AND
HAROLD MASURSKY

> "These earthly godfathers of heaven's lights
> that give a name to every fixed star...."
>
> Shakespeare, *Love's Labour's Lost*

4.1. EARLY NOMENCLATURE: PRETELESCOPIC PERIOD

The godfathers of Earth have indeed been busy naming "heaven's lights" and features on them, particularly during the past sixty years. During this period, the official godfather has been the International Astronomical Union (IAU), which, since its founding in 1919, has been the governing body for all planetary nomenclature. Under its direction, more than forty-five hundred names have been adopted for planets, satellites, and features on these bodies, discriminated by twentieth-century technology; several thousand other names have been given to asteroids and comets. Why do people name things? Why are names for planetary features important? And is such proliferation of planetary names really necessary?

Humankind has been naming the stars and the "wandering stars" – the planets – for thousands of years. Planets were recognized very early as being different from other stars, partly by their movement but also by differences in color or magnitude compared to the relatively immovable stars. Names were applied originally, as all names are, to identify the objects uniquely. Interest in the planets and the development of astronomical information appears to have been worldwide and to have begun at a very early period in man's history.

Names given to the planets by early peoples were usually those of a god or goddess, or they described an attribute of the god or of the "star" itself. Many of the names have survived to the present time, notably as designations for days of the week. Mars, for instance, was known as Tiu (from which we derive *Tuesday*) in Old English. The Roman name for the planet was Mars, their god of war; later its two satellites were named Phobos (Fear) and Deimos (Terror) after Mars' servants, the mythological horses that pull his chariot. The French word for Tuesday, *Mardi*, is derived from Mars. In ancient Egypt, Mars was called Harmakis, or Her deshur ("the red one," from its color). The ancient Greeks associated the red planet with their bloody god of war,

Ares. The Chaldeans and Babylonians called it Nirgal, for the Babylonian warlike hero.

Plurality of nomenclature did not cause conflict or confusion as long as the body could be identified unequivocally by its position in the sky and relative brightness or color: The Chaldeans and Egyptians presumably learned to equate Nirgal with Her deshur; the Greeks and Romans had no trouble interchanging Ares and Mars. With the invention of the telescope in 1608, however, astronomers from many countries began studying the Moon and other planetary bodies. Each applied his own names to the features he observed on each body, so several different systems of nomenclature were developed to identify albedo or topographic features visible by telescope from Earth. Several systems were formulated for naming albedo features on extraterrestrial bodies, but the most extensive, and chaotic, nomenclature was that developed for the Moon.

4.2. MOON: 1640–1977

The first maps of the Moon to show nomenclature were drawn in the 1640s by the Polish astronomer Hevelius (Johann Höwelcke), who named albedo regions for places in classical geography, and the Belgian astronomer Langrenus (Michel Florent van Langren), who named craters for persons or places. In 1651 the Italian astronomer Giovanni Riccioli named dark areas "maria" – seas – and gave them symbolic names like Mare Serenitatis, meaning "sea of serenity"; he considered light areas to be continents. Crater names were chosen to honor deceased, internationally famous persons, like Aristarchus. Riccioli's system eventually became the standard from which future lunar nomenclature developed.

During the eighteenth and nineteenth centuries, Riccioli's nomenclature was expanded and changed by the astronomers who followed him. In the 1780s the German astronomer Johann Schröter named additional features and initiated the convention of applying Roman or Greek letters to small formations near a larger, previously named formation; however, his assignment of letters was haphazard. The German astronomer Johann Mädler, in the 1820s, systematized Schröter's Greek and Roman letters: Roman capital letters were applied to small satellite craters, and Greek letters were given to formations, such other as rilles, crests, or mountains; Reiner A and Reiner Gamma are examples. In the 1870s, Johann Schmidt (German) and Edmund Neison (English) named many additional features, following Mädler's scheme. In the early 1900s, Samuel Saunder (English) and Julius Franz (German) prepared catalogs of the previously named features, changing some designations and naming some additional features.

By the early 1900s, many of the prominent features on the Moon's nearside were known by at least three names. The need for a uniform lunar nomenclature was brought before Great Britain's Royal Astronomical Society and, through the society, to the International Association of Academies. This group appointed a committee to collate the nomenclatures developed by Mädler, Schmidt, and Neison. The British astronomer Mary A. Blagg completed this task (Blagg, 1913).

When the IAU was founded in 1919, one of its first actions was to form a committee to regularize lunar nomenclature; Blagg, Karl Müller, and several other members of the earlier committee were appointed to the new group. This committee selected one name for each feature on Blagg's collated list, and added a few names that were either new or taken from other schemes. The resulting compendium, *Named Lunar Formations,* was presented to Commission 17 in 1932 and adopted by the General Assembly of the IAU in 1935 as the official list of 572 approved names (Blagg and Müller, 1935).

The Blagg and Müller scheme set the style not only for lunar nomenclature, but for future nomenclatural themes that would be adopted by the IAU for features on other planets and satellites in the solar system. (Small bodies – asteroids and comets – utilize a different system of nomenclature.) Henceforth, craters on planets and satellites would be named for famous personages (deceased, if human). Other features would usually be given a double designation, one of which would be a latinized definitive term, such as *mare* (sea), or *vallis* (valley). Nine terms are assigned to lunar features in the Blagg and Müller catalog; of these, only three were adopted for use on other planets: *crater, rupes,* and *vallis.* The second part of each name would be assigned from themes individualized to each planet or satellite.

As space exploration expanded and images of other planetary surfaces revealed features unlike those seen on Earth or the Moon, additional Latin terms for geomorphic and albedo features were proposed (Table 4.1). In addition, categories for naming noncrater forms have been chosen for features on all twenty-five planets and satellites, other than the Moon, for which images have been obtained during the past twenty years (Table 4.2). The connotative nature of the theme chosen for each body and the definitive nature of the latinized term make it possible to determine, from its name, the location (by body) and geomorphic style of each named feature.

The IAU's Nomenclature and Cartography Committee of the Lunar Surface continued to name craters after a few preeminent deceased scientists during the 1940s and early 1950s. Then, in the mid-1950s, Gerard P. Kuiper, a noted Dutch astronomer who had emigrated to the United States, initiated a project to map the Moon in much greater detail. Kuiper started his work at the University of Chicago but moved in the late 1950s to Tucson, Arizona, where he founded the Lunar and Planetary Laboratory in 1960. During the next eight years, Kuiper and two associates, David W. G. Arthur and Ewen A. Whitaker, completely updated and reorganized the lunar nomenclature. At the Twelfth General Assembly of the IAU in Hamburg in 1964, *The System of Lunar Craters,* a catalog and a map in four quadrants (Arthur et al., 1963, 1964, 1965, 1966) was adopted as the official IAU document to show positions and list selenographic information of all approved lunar nomenclature, including all small cra-

Table 4.1. *Definitive feature terms used in planetary nomenclature.*

Feature	Description
Catena, catenae	Chain of craters
(Cavus) cavi[a]	Hollows, irregular depressions
Chaos[b]	Distinctive area of broken terrain
Chasma, chasmata	Canyon
Collis, colles	Small hill or knob
Corona, coronae	Ovoid-shaped feature
Crater, cratera	Crater
Dorsum, dorsa	Ridge
Eruptive center	Eruptive center
Facula, faculae	Bright spot
Flexus, flexūs	Cuspate linear feature
Fluctus, fluctūs	Flow terrain
Fossa, fossae	Long, narrow, shallow depression
Labes, labēs	Landslide
Labyrinthus, labyrinthi	Intersecting valley complex
Lacus, lacūs	"Lake"; small irregular dark area
Landing site name	Feature named on Apollo map or report
Large ringed feature	Large ringed feature
Linea, lineae	Elongate marking
Macula, maculae	Dark spot
Mare, maria[c]	"Sea"; large low plain
Mensa, mensae	Mesa, flat-topped elevation
Mons, montes	Mountain
Oceanus, oceani[c]	"Ocean"; very large, low plain
Palus, paludes[c]	"Swamp"; irregular low plain
Patera, paterae	Shallow crater, scalloped or complex edge
Planitia, planitiae	Low plain
Planum, plana	Plateau or high plain
Promontorium, promontoria[c]	"Peninsula"; elongate upland area
Regio, regiones	Region
Rima, rimae[c]	Fissure
Rupes, rupēs	Scarp
Scopulus, scopuli	Lobate or irregular scarp
Sinus, Sinūs[c]	"Bay"
Sulcus, sulci	Subparallel furrows and ridges
Terra, terrae	Extensive upland area
Tessera, tesserae	"Tile"; polygonal ground
Tholus, tholi	Small domical mountain or hill
(Unda), undae[a]	Dunes
Vallis, valles	Sinuous valley or linear depression
Vastitas, vastitates	Widespread lowland area

[a]Used only in plural form. [b]Indeclinable term; plural not used. [c]Used only on the Moon.

ters identified by letter and name. Named features are indicated by outline on a map that shows coordinates but no features. The *Rectified Lunar Atlas* (Whitaker et al., 1963) was also adopted at the same meeting. In the *Atlas,* photographs of the Moon's nearside, divided into thirty sections, were projected on a globe to remove the effects of foreshortening on the limbs. Nomenclature is shown overlain on a shaded relief rendition of the photographic map.

Table 4.2. *Categories for naming features on planets and satellites.*

MOON

Craters, catenae, dorsa, rimae	Famous deceased scientists, scholars, artists; common first names for small craters
Lacūs, maria, paludes, sinūs	Latin adjectives describing weather conditions
Montes	Terrestrial mountain ranges
Valles	Nearby features

MERCURY

Craters	Famous deceased artists, musicians, painters, and authors
Montes	Caloris, from word for "hot"
Planitiae	Names for Mercury in various languages
Rupes	Ships of discovery or scientific expeditions
Valles	Radio telescope facilities

VENUS

Chasmata	Goddesses of hunt; moon goddess
Coronae	Fertility goddesses
Craters	Famous women
Dorsa	Sky goddesses
Fluctūs	Goddesses, miscellaneous
Lineae	Goddesses of war
Montes	Goddesses, miscellaneous (also one radar scientist)
Paterae	Famous women
Planitiae	Mythological heroines
Planum (1 only)	Goddess of prosperity
Regiones	Giantesses and titanesses (also Greek alphanumeric)
Rupēs	Goddesses of hearth and home
Tesserae	Goddesses of fate or fortune
Terrae	Goddesses of love

MARS

Large craters	Deceased scientists who have contributed to the study of Mars
Small craters	Villages of the world (less than 100,000 population, U.N. Yearbook)
Large valles	Name for Mars/star in various languages
Small valles	Classical or modern rivers
Other features	From nearest named albedo feature on Schiaparelli or Antoniadi maps

MARTIAN SATELLITES

Deimos	Authors who wrote about satellites
Phobos	Scientists who helped discover Phobos

JOVIAN SATELLITES

Callisto

Large ringed features	Homes of the gods and heroes
Craters	Heroes and heroines from northern myths

Table 4.2. (*cont.*)

Europa

Craters	European (Celtic) gods and heroes
Lineae	People associated with the Europa myth
Faculae	Egyptian places
Flexūs, maculae	Places associated with the Europa myth

Ganymede

Craters	Gods and heroes of ancient (Fertile Crescent) people
Sulci	Places associated with myths of ancient people
Facule	Places associated with Egyptian myths
Fossae	Gods (or principals) of ancient (Fertile Crescent) people

Io

Active eruptive centers	Fire, sun, and thunder gods and heroes
Paterae	Volcanic gods and goddesses, mythical blacksmiths; fire, Sun, thunder gods and heroes
Montes	Mountains associated with Io myth
Plana, Regiones	Places associated with Io myth
Tholi	People associated with Io myth

SATURNIAN SATELLITES

Mimas	People and places from Malory's *Le Morte d'Arthur* legends (Baines translation)
Enceladus	People and places from Burton's *Arabian Nights*
Tethys	People and places from Homer's *Odyssey*
Dione	People and places from Virgil's *Aeneid*
Rhea	People and places from creation myths
Hyperion	Sun and Moon deities
Iapetus	People and places from Sayers' translation of *Song of Roland*
Epimetheus	Castor and Pollux (twins) myth
Janus	Castor and Pollux (twins) myth

URANIAN SATELLITES

Puck	Mischievous spirits
Miranda	Humans from Shakespeare's *The Tempest*
Ariel	Bright (good) spirits; individuals and classes, worldwide mythologies
Umbriel	Dark (evil) spirits, worldwide mythologies
Titania	Minor Shakespearean female characters
Oberon	Shakespearean heroes

NEPTUNIAN SATELLITES

Triton	Aquatic gods, goddesses, places
Nereid	Theme to be decided

Six as yet unnamed satellites

In the meantime the launch of Sputnik in 1957 catalyzed the American and Soviet race to explore and map the Moon and other extraterrestrial bodies. The Lunar and Planetary Laboratory's effort to map the Moon was serendipitous to the announcement by President John F. Kennedy in 1962 that the United States intended to land men on the Moon within that decade.

With each succeeding Soviet and American mission to the Moon, additional lunar features that required names were recognized. As a result of the Soviet-sponsored Zond and Luna missions and the American-sponsored Ranger, Lunar Orbiter, and Apollo missions, the IAU was presented with several problems concerning contested or duplicated names. For example, the Soviets had chosen Russian names for most of the features on the lunar farside because they were the first to image the Moon's hidden hemisphere, but names of features on the nearside were international in representation. The need for an objective group to arbitrate this departure from convention and to referee the naming process increased with each succeeding year and mission. For this purpose a Working Group on Lunar Nomenclature was established in 1967 by Commission 17 of the IAU; the Group was chaired by Donald Menzel and was asked to determine which names should be adopted and what additional guidelines for planetary nomenclature should be developed.

At the Brighton meeting of the IAU in 1970, more than five hundred new names (chiefly for farside features) were adopted. Menzel had developed a representative and international list of candidate names by requesting nominations from worldwide national academies of science. Soviet and American names dominate the final list (each country is represented by more than one hundred honorees) because these countries were the first to photograph the Moon's farside. However, the other three hundred names represent persons from more than thirty other countries. In this way the convention initiated by Riccioli of according international representation to named features became the rule.

The IAU also agreed, after heated debate at the Brighton General Assembly, to expand letter-designated small craters to the farside and to name craters for three deceased American astronauts (Chaffee, Grissom, and White) and for the six living American astronauts of the Apollo 8 and 11 crews who were the first humans to reach or orbit the Moon. An equal number of living and deceased Soviet cosmonauts were similarly honored. The lettered-crater scheme was expanded so that farside and nearside nomenclature would be conformable. The Apollo 8 and 11 missions, and the various cosmonaut discoveries, were considered to be sufficiently important to warrant departure from the convention of honoring only deceased persons. Decisions made at the Brighton meeting and a complete list of named features adopted at that meeting are contained in an article printed in *Space Science Reviews* (Menzel et al., 1971).

The Working Group on Lunar Nomenclature continued to function during the next trienniel period. In 1973 the Group was reconstituted as a Task Group that, for the next six years, added many names to features discriminated by Apollo photographs and shown on maps generated from them. Included were a few operational names (IAU, 1977) that had been applied informally to features seen by Apollo astronauts from orbit and a larger set of names for features visited by them during mission traverses (IAU, 1977). Figure 4.1 shows a map of part of the lunar farside, exemplifying present lunar nomenclature.

4.3. MARS: MID-SEVENTEENTH CENTURY TO 1976

At the same time that the Apollo missions to the Moon were monopolizing most public and scientific interest, some planetary scientists were turning their attention to Mars. In 1965 and 1969 the American Mariner 4, 6, and 7 spacecraft successfully imaged small areas of the Martian southern hemisphere. Photographs returned by these missions showed a heavily cratered surface, similar to that of the lunar highlands. No apparent correlation could be seen between the albedo features mapped from telescopic observations and the geomorphic features discriminated by Mariner cameras. Clearly, the existing nomenclature, developed for albedo features, would be inadequate for naming the many topographic features that the forthcoming Mariner 9 mission was expected to image when it obtained planetwide coverage in 1971.

A nomenclature had been developed earlier to identify albedo features on Mars, but it was as disparate as that of the early lunar nomenclature. Since the mid–seventeenth century, when the Dutch astronomer Christiaan Huygens drew the first map of Mars, astronomers had been drawing and, later, photographing the tenuous dark and light albedo features they recognized in telescopic observations; each had devised his own system of nomenclature. The German astronomers Wilhelm Beer and Johann Mädler, who drew the first complete map of Mars in 1830, designated albedo areas by letter. Twenty years later the Italian astronomer P. A. Secchi named individual areas for explorers (Cabot, Columbus, Cook, etc.). In 1867 the English astronomer R. A. Proctor mapped the light and dark areas of Mars in greater detail, calling the light areas continents, lands, and islands, and the dark areas seas, straits, or lakes. Proctor named these albedo features for past and current astronomers who had contributed to the study of Mars. (On a later map published in 1888, symbolic names, such as Phaethontis, or "storm land," were added; these names are similar to names applied much earlier to lunar maria features.) In the 1860s the French astronomer Camille Flammarion and the English astronomer Nathan Green also drew maps that were strongly influenced by Proctor's early map; however, they differed enough in detail that Proctor could not claim preeminence for his nomenclature.

In 1877 the Italian astronomer G. V. Schiaparelli was able to delineate features with much greater clarity than had Proctor or other predecessors. Schiaparelli could subdivide Proctor's four large "continents" into many

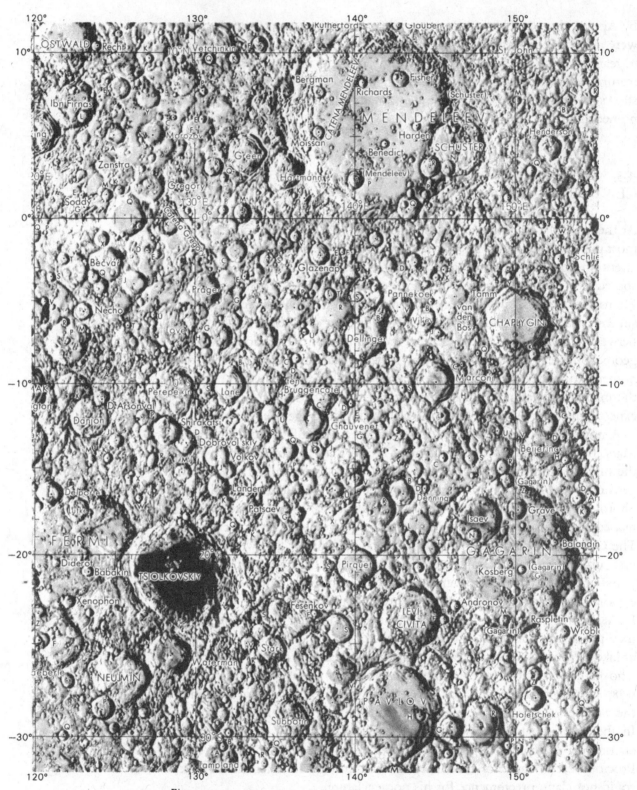

Figure 4.1
Lunar nomenclature, as shown on part of Relief and Surface Markings of Lunar Far Side *(U.S. Geol. Survey Misc. Inv. Ser. Map I-1218-A).*

smaller, clearly defined "island" areas; in addition, several of Proctor's dark "sea" areas appeared to Schiaparelli to be much smaller in size. Proctor's nomenclature was too general and imprecise for the many additional features Schiaparelli observed, so Schiaparelli developed a new scheme for naming the newly defined areas. In his system he chose familiar names that would be easily remembered by members of the astronomical community: names taken from classical mythology and its related geography that, from earliest times, had provided the names of planets, constellations, and stars. As he gazed through his telescope, he imagined he saw, in the albedo features of Mars, the classical world centered on the Mediterranean Sea; thus, he could trace on his Martian map the wanderings of Odysseus and other mythological heroes of classical literary fame. His first map, drawn in 1877, named features in the southern hemisphere of Mars; in 1879 he enlarged the map to include some features in the northern hemisphere and also named several thin, dark streaks ("canali") for classical rivers (Schiaparelli, 1877–8, 1880–1; Blunck, 1977, pp. 165–7).

Schiaparelli's nomenclature was accepted by many European astronomers but denounced by others, notably Otto von Struve of the Pulkowa Observatory. Struve preferred Proctor's system of naming the features for contemporary astronomers. A spirited debate ensued; the Europeans generally sided with Schiaparelli, but the English and American astronomers mostly supported Proctor. The division was not entirely along national or ethnic lines, however. The English astronomer H. Sadler pointed out that naming the features for contemporary astronomers was dangerous "inasmuch as it anticipates the verdict of posterity." J. B. Lindsay, also English, used alphabetical designations or the names of male mythological characters. The American astronomer Percival Lowell enthusiastically supported Schiaparelli, probably because of their shared interest in the dark streaks Lowell had labeled "canali." On the other hand, the German astronomer Albert Marth preferred a system that used large and small letters to identify the features and subfeatures. By the end of the nineteenth century, however, Schiaparelli's names for the Martian features prevailed and were accepted by most other scientists.

Many of the questions raised at that time concerning Martian nomenclature have been shown to be valid; several of the alternative naming schemes were utilized later to name planetary features, especially when a nomenclature was in its early stages. Sadler's doubt about assigning names of living persons has been shown to be well founded; the IAU was to rule that a person must be deceased for at least three years before his name could be assigned to a planetary feature. Mythological characters and places, used sparingly on the Moon and extensively on Mars, became the theme for naming features on the Outer Planets. Greek letters were used in early nomenclature of Venus to identify areas of high reflectivity, and a scheme similar to Marth's was used to identify small craters on maps of Mars drawn from Mariner 9 data. In both cases the system of identifying features by letter was dropped as the nomenclature became more complex.

During the first third of the twentieth century Percival Lowell (founder of the observatory at Flagstaff, Arizona) and the Italian-born astronomer E. M. Antoniadi (working at Juvisy, France) were the most prolific mappers of Mars. Of these, Antoniadi's work was more widely accepted; during that period, he (and others) greatly expanded the Schiaparelli system until more than four hundred names were in general use. Antoniadi also introduced a few additional variations on the classical mythological theme and included the names of constellations and stars, such as Mare Serpentis, and of the color of various areas, such as Claritas. Most of the names on his map (Antoniadi, 1930) were taken from Schiaparelli's map or were additions that followed Schiaparelli's system. The Antoniadi map and its predecessor map by Schiaparelli are the bases on which modern Martian nomenclature has been developed. A book by the German historian Jürgen Blunck describes in detail the Schiaparelli–Antoniadi nomenclature as well as the modern adaptations of their nomenclature (Blunck, 1977, 1982).

By 1948, however, the Martian nomenclature had slipped back into a chaotic state, as various astronomers changed or replaced names for no apparent reason. Once again the IAU was asked to regularize nomenclature, this time for Mars. A subcommission appointed in 1952 and chaired by Georges Fournier pared the list of 558 names then in use to 404. After Fournier's death in 1954 a new committee, chaired by Auduoin Dollfus, reduced the number of names proposed for approval to 128; 105 of these names were derived from Schiaparelli's map and sixteen from Antoniadi's. These names were adopted by the IAU General Assembly at its Moscow meeting in 1958 (IAU, 1960), but most astronomers continued to use a much larger list of names derived from the Antoniadi and Schiaparelli maps.

Then in 1967 an ad hoc committee chaired by Gerard Kuiper was established to develop a plan for naming geomorphic features imaged by the Mariner cameras. This committee suggested that large craters on Mars, like those on the Moon, be named for deceased scientists or other notable persons who had contributed to the study of Mars. Later, at the Brighton meeting of the IAU in 1970, Commission 16 formed a Working Group for Martian Nomenclature (WGMN); Gerard de Vaucouleurs was named chairman. The committee was asked to address the specific problem of expanding the Martian nomenclature to accommodate the large number of features that Mariner 9 was expected to image. A few informal names for craters (including the name Mariner for the first large crater imaged by the Mariner spacecraft) were chosen by this group, and a coordinate system was established and utilized on an albedo map of Mars used for planning the Mariner 9 mission (JPL, 1971).

At the IAU meeting in 1973 at Sydney, Australia, after Mariner 9 had successfully completed its mission, the WGMN presented a plan for Martian nomenclature that was patterned after the lunar system of nomenclature yet preserved the familiar names of the early Martian albedo nomenclature. Albedo names from the 1958 list were adopted for thirty provinces; thirteen descriptive Latin or Greek terms were chosen that described the geomorphic appearance of features; and 179 large craters and almost one hundred other prominent features were given specific names. Of the thirteen terms adopted, only three (*mons, vallis,* and *rupes*) had been used previously on the Moon. This expansion of the nomenclature marks the change in emphasis from astronomical to planetary scientific studies.

Large craters were named for deceased scientists and explorers or science fiction writers whose work was associated with Mars. Most of the names chosen for Martian craters had also been assigned earlier to lunar craters; the duplication of names proved confusing and led to the rule, adopted later, that names would not be duplicated except in extraordinary circumstances. Large channels were named for the word *Mars* in various languages. Some of the ancient names of Mars adopted at this time are discussed in the introduction to this chapter; other adopted names include Al-Qahira (Arabic), Ma'adim (Hebrew), and Kasei (Japanese). Somewhat later, the IAU decided to choose a different category for small channels, now named for classical and modern rivers of the world, and for small craters, now named for villages of the world. Originally, municipalities of less than one hundred thousand population were considered "villages"; however, many "villages" rapidly became "cities," and an informal cutoff of about twenty thousand population is now used.

All other features are named by combining an albedo name, derived from Antoniadi's map of Mars, with a Latin or Greek term that describes the topographic form of the feature. In most cases the gender, number, and order of the Latin or Greek term is forced into agreement with that of the albedo term used on Antoniadi's map: Arsia Mons, from the albedo name Arsia Silva; Tithonium Chasma, from the albedo term Tithonius Lacus. Table 4.2 lists the categories used to name features on Mars; Figure 4.2 is an example of Martian nomenclature.

Names for features on the Martian satellites Phobos and Deimos were also assigned at Sydney. Craters on Phobos are named for scientists who have studied the Martian satellites; Deimos' craters are named for authors who have written about the satellites.

Since the Sydney meeting in 1973, Martian nomenclature has been expanded to include more than one thousand adopted names for twenty-two definitive terms (WGPSN, 1986; Table 4.1). Such proliferation became necessary as investigators analyzed and wrote about the Martian surface, as it was revealed in increasing detail, first by the Mariner images and later by images of much higher resolution obtained by the Viking mission. Additional guidelines and constraints for planetary nomenclature were also developed, based mostly on lunar and Martian experience. Later, as missions to other planets revealed surface features unlike those on the Moon and Mars, the rules and the number of feature types were expanded to accommodate the new situations. These and other refinements to the rules, added later, are listed at the end of this chapter.

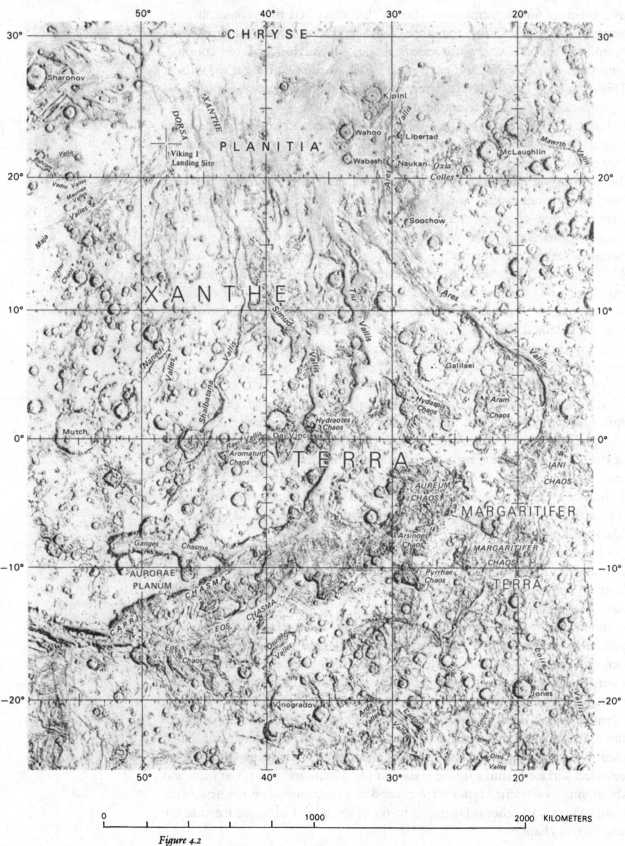

Figure 4.2
Martian nomenclature, as shown on part of Western Region of Mars *(U.S. Geol. Survey Misc. Inv. Ser. Map I-1618, sheet 2).*

4.4. IAU TASK AND WORKING GROUP FORMAT CHANGE – 1973

In addition to the extensive work done on Martian nomenclature at the Sydney meeting, a basic change was adopted in the administrative format of the nomenclature committees. The Lunar and Martian Working Groups were reformatted as Task Groups, and a Working Group for Planetary System Nomenclature (WGPSN) was formed to oversee the work of the Task Groups and to formulate policy on all matters concerning planetary nomenclature. Peter Millman of the National Research Council of Canada was named chairman of the WGPSN, and Bradford Smith of the University of Arizona replaced de Vaucouleurs as chairman of the Mars Task Group. Nomenclature Task Groups were also appointed for Mercury, Venus, and the Outer Planets; David Morrison (University of Hawaii), Gordon Pettengill (Massachusetts Institute of Technology), and Tobias Owen (State University of New York at Stony Brook) were named as the respective chairmen of these groups. Membership of the various Task Groups and the WGPSN has remained much the same since their formation, except for a few additions and, in some instances, a change in chairmanship. V. V. Shevchenko (Sternberg State Astronomical Institute, Moscow) became chairman of the Lunar Task Group when Donald Menzel died in 1974, M. Ya. Marov (Institute of Applied Mathematics, Moscow) replaced Gordon Pettengill as chairman of the Venus Task Group in 1984, and Harold Masursky (U.S. Geological Survey, Flagstaff) replaced Peter Millman as chair of the WGPSN in 1983. Present makeup of the Working and Task Groups is shown in Table 4.3.

4.5. MERCURY: 1974–6

The newly formed rules for planetary nomenclature were put into immediate use by the Mercury Task Group as it prepared to name features imaged by the Mariner 10 spacecraft in 1974 and 1975. Like those of Mars, albedo features on Mercury had been mapped and named by Antoniadi (Antoniadi, 1934), and the Task Group planned originally to use names from Antoniadi's map, as the Mars Task Group had done. Individual maps of the 1:5,000,000-scale, shaded relief series of Mercury were assigned albedo names in 1976 (IAU, 1977) by using a new albedo map drawn in 1972 (Murray et al., 1972) as a base for the Antoniadi names. However, by the time the shaded relief maps of the quadrangles had been drawn, the Task Group had abandoned this system in favor of one that would honor a more diverse sector of humankind. Craters are named for eminent deceased authors, artists, and musicians. Low plains (planitiae) are given the names of Mercury in various languages, a category similar to that chosen for naming Martian large channels. Scarps (rupēs) are named for ships of discovery, ridges (dorsa) for astronomers who drew albedo maps of Mercury, and valleys (valles) are named for radar

Table 4.3. *Membership in IAU Task Groups and the Working Group for Planetary System Nomenclature.*

WORKING GROUP FOR PLANETARY SYSTEM NOMENCLATURE

(A committee of the Executive Committee)

PRESIDENT: H. Masursky, U.S.A.

MEMBERS:		CONSULTANTS:	
K. Aksnes	Norway	J. M. Boyce	U.S.A.
G. E. Hunt	U.K.	G. A. Burba	USSR
M. Ya. Marov	USSR	A. M. Komkov	USSR
P. M. Millman	Canada	W. E. Brunk	U.S.A.
D. Morrison	U.S.A.		
T. C. Owen	U.S.A.		
V. V. Shevchenko	USSR		
B. A. Smith	U.S.A.		
V. G. Tejfel	USSR		

TASK GROUP FOR LUNAR NOMENCLATURE

V. V. Shevchenko (Chair)	USSR
A. Dollfus	France
F. El-Baz	U.S.A.
H. Masursky	U.S.A.
P. M. Millman	Canada
S. K. Runcorn	U.K.
E. A. Whitaker	U.S.A.

TASK GROUP FOR MERCURY NOMENCLATURE

D. Morrison (Chair)	U.S.A.
D. B. Campbell	U.S.A.
M. E. Davies	U.S.A.
A. Dollfus	France
N. P. Erpylev	USSR
J. E. Guest	U.K.

TASK GROUP FOR VENUS NOMENCLATURE

M. Ya. Marov (Chair)	USSR
A. T. Basilevsky	USSR
D. B. Campbell	U.S.A.
R. M. Goldstein	U.S.A.
R. F. Jurgens	U.S.A.
H. Masursky	U.S.A.
G. H. Pettengill	U.S.A.
Y. F. Tjuflin	USSR

TASK GROUP FOR MARS NOMENCLATURE

B. A. Smith (Chair)	U.S.A.
A. Dollfus	France
M. Ya. Marov	USSR
D. Ya. Martynov	USSR
H. Masursky	U.S.A.
S. Miyamoto	Japan
C. Sagan	U.S.A.

TASK GROUP FOR OUTER SOLAR SYSTEM NOMENCLATURE

T. C. Owen (Chair)	U.S.A.
K. Aksnes	Norway
A. T. Basilevsky	USSR
R. Beebe	U.S.A.
M. S. Bobrov	USSR
A. Brahic	France
M. E. Davies	U.S.A.
N. P. Erpylev	USSR
H. Masursky	U.S.A.
B. A. Smith	U.S.A.
V. G. Tejfel	USSR

TASK GROUP FOR SURFACE FEATURES ON ASTEROIDS AND COMETS

M. Fulchignoni (Chair)	Italy
J. Veverka	U.S.A.
A. Brahic	France
D. Morrison	U.S.A.
T. Gombosi	Hungary
L. Ksanfomaliti	USSR
G. Lupishko	USSR
Y. C. Chang	China
S. Isobe	Japan

observatories. One large planitia and the rough (montes) region bordering it are named Caloris because that part of Mercury is the hottest region of the planet (Figure 4.3). Other departures from standardized categories include a crater named for Kuiper (a scientist) and a small crater named Hun Kal, the Mayan word for *twenty,* because it lies on the twentieth meridian and sets the coordinate system for Mercury. An example of Mercurian nomenclature is shown in Figure 4.3.

4.6. BETWEEN MISSIONS, 1976–9

The period between the General Assemblies of 1976 and 1979 was a busy time for the various task groups and for the WGPSN. Additional features were named on the 1:5,000,000-scale maps of Mercury. The Lunar Task Group named a number of features on high-resolution Apollo maps, including small craters that were given male and female first names taken from worldwide cultures. Names for features on the 1:1,000,000-scale, Viking 1 landing site map were adopted in 1976, but names for features on the second Viking landing site maps and for the high-resolution map of the first landing site were debated for another two years and were not adopted until the 1979 meeting. The Viking landings of 1976 were intended to be part of the American bicentennial celebration, and the Mars Task Group had planned to name thirteen craters on the Viking 1 high-resolution landing site map for the most important town in each of the original thirteen American colonies. Although the Working Group approved the bicentennial theme, it insisted that an IAU rule, requiring that all maps be international in scope, must be followed. After a year's spirited debate, fifteen additional craters were named for worldwide ports that traded with the colonies in 1776. Craters on the high-resolution map of the second Viking landing site were named for radar and launch facilities that in 1976 are part of modern space missions, including Viking.

Stimulated by launches of the two Voyager spacecraft in 1977 and the Pioneer Venus spacecraft in 1978, the Outer Planets and Venus Task Groups began compiling lists of names related to categories they planned to use for features discriminated by Voyager's cameras and Pioneer Venus' radar.

4.7. VENUS: 1960–88

Venus, too, had an informal nomenclature that had been developed to identify features recognized in Earth-based radar images. Monostatic and pulsed Earth-based radar systems that were able to detect echoes from the surface of Venus had been developed in the early 1960s. Low-resolution images, obtained for fifteen years prior to the Pioneer Venus mission by scientists using the Arecibo (Puerto Rico) and Goldstone (California) radar telescopes, showed some anomalously bright areas and other dark areas of low reflectivity. As early as

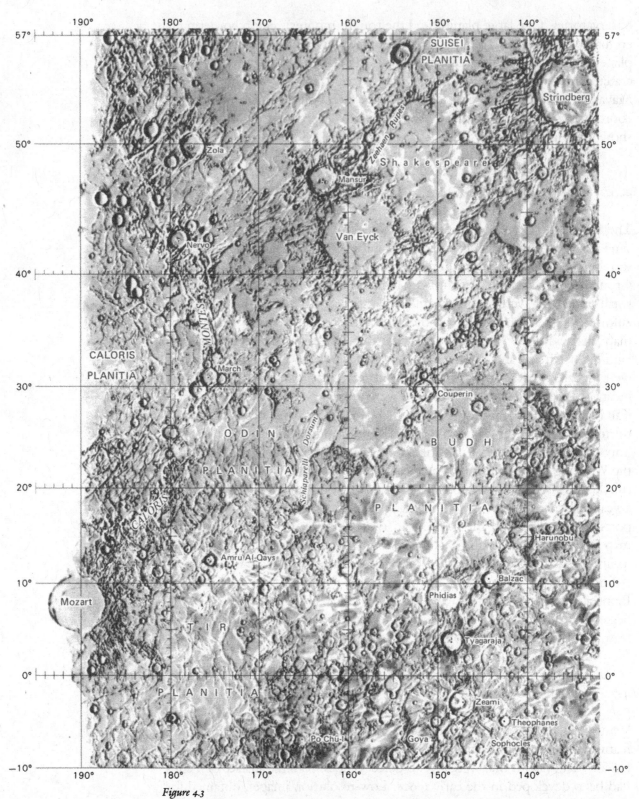

Figure 4.3
Mercurian nomenclature, as shown in the western equatorial section of Shaded Relief Map of Mercury *(U.S. Geol. Survey Misc. Inv. Ser. Map I-1149).*

1964, bright areas in the Goldstone images were named Alpha, Beta, etc.; in 1967 Arecibo scientists named several bright areas for famous radio and radar scientists, such as Hertz and Maxwell. Other scientists used Roman letters or numbers to define the radar-bright areas. These preliminary systems of nomenclature were similar to schemes used earlier for lunar and Martian nomenclature. However, none of the systems seemed, to the Venus Task Group, to provide a sufficiently large bank of names to accommodate the number of topographic features that were being recognized in more sophisticated Earth-based images, as well as the much larger number that was expected to be elucidated by the Pioneer Venus radar images and radar altimetry. Therefore, the Task Group chose a theme in keeping with the age-old feminine mystique associated with Venus: features would be named for females, both mythological and real, from the mythologies and histories of cultures throughout the world (Figure 4.4). Circular, craterlike forms would be named for notable, historical women, and other features would bear the names of mythological or legendary goddesses and heroines.

As the Pioneer Venus radar data became available, names were applied to prominent features, and feature classes were assigned (Figure 4.4). Three names that were well established in the radar literature were retained from the early nomenclature: Maxwell became Maxwell Montes (the only male name on Venus); Alpha and Beta became Alpha Regio and Beta Regio, two of the albedo-bright areas of lesser topographic relief above the widespread Venusian plains. Three highland areas the size of terrestrial continents were named for counterparts of Venus, the Roman goddess of love: Ishtar Terra, Aphrodite Terra, and Lada Terra, for the Babylonian, Greek, and Russian love goddesses. One high plateau area was named Lakshmi Planum to honor the Indian goddess of prosperity and fortune. Lowland plains are named for mythological heroines: Helen Planitia honors the Greek woman whose face "launched a thousand ships," and Sedna Planitia is named for a heroine of Eskimo mythology. Linear canyons (chasmata) on the Venusian surface are named for goddesses of the hunt, or Moon goddesses, because mythological personages such as Artemis (Greek) or Diana (Roman) are often assigned both attributes. The names of hearth goddesses were applied to areas of abrupt topographic change (cliffs or scarps); for example, Vesta Rupes is named for the Roman hearth goddess.

The names of notable deceased women are given to circular or irregular, bowl-shaped features. Those near or at the summits of highland areas were designated paterae (Table 4.1) and named for classical women (Cleopatra Patera), whereas those in plains areas that are more rounded in shape were named for modern women (Lise Meitner) and designated craters. However, this distinction between paterae and craters has been challenged. It is very difficult to determine whether a feature on Venus should be designated as a patera or a crater, because resolution is still too poor to discriminate crater ejecta material, which is used to differentiate impact craters from volcanic features on other planets or satellites.

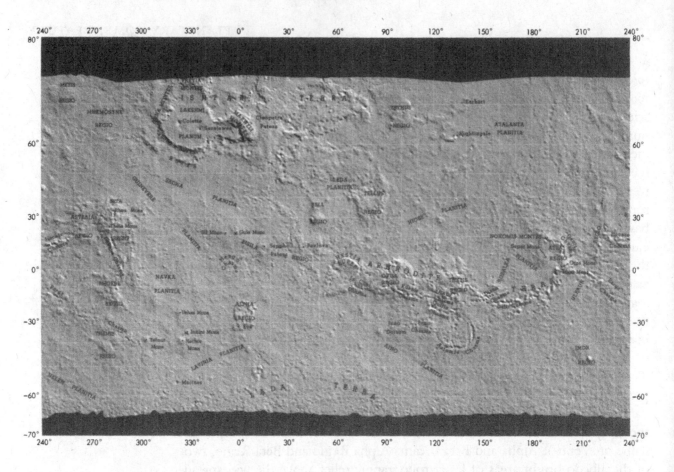

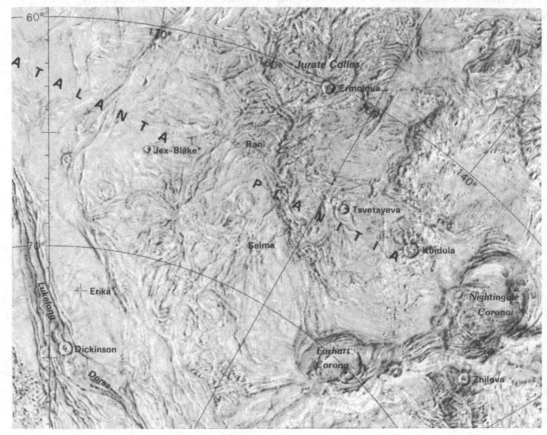

B

Other features with little or no relief were assigned to feature classes that are usually reserved for albedo features; regio and linea are in this category. Aside from Alpha Regio and Beta Regio, these areas are named for titanesses or giantesses; Metis Regio is an example. Radar-bright linear features that appear only in reflectivity images were named for goddesses and heroines of war and were assigned the term "linea"; the term was being suggested concurrently for albedo features on the Jovian satellite Europa. Vihansa Linea, named for a Teutonic goddess, is an example.

4.8. JOVIAN SATELLITES: 1977–80

Although the Voyager spacecraft were launched in 1977, almost a year before Pioneer Venus launch, the two missions arrived at their destinations at about the same time. In 1977 and 1978 the Outer Planets Task Group was as active as the Venus Task Group, developing a theme and compiling a bank of names for the features on the Jovian satellites that Voyager would image.

A few informal names had been applied to features in Jupiter's atmosphere ("the great Red Spot," and so-called belts and zones), but the satellites were much too small and poorly observed before Voyager to have acquired names for their albedo features. The four largest satellites of Jupiter were discovered by Galileo in 1610, shortly after the first telescope was built; they have been called the Galilean satellites since that time, although Galileo did not give them their present names. Galileo called them the "Medicean stars" to honor his patron, Cosimo II de' Medici, but astronomers of that day were unwilling to accept such a chauvinistic approach to nomenclature. They may also have questioned the propriety of honoring a living person whose place in posterity was uncertain. The satellites were named in 1610 for mythological characters beloved by Jupiter – Io, Ganymede, Europa, Callisto – by the German astronomer Marius (Simon Mayer). Galileo and Marius discovered the satellites at about the same time, but neither was aware of the other's work. A bitter controversy arose to determine who would be given credit for the discovery and what the satellites would be named. Eventually the scientific community recognized Galileo as the discoverer but accepted Marius' names.

The Outer Planets Task Group had decided several years before Voyager's arrival at Jupiter that the mythological theme originated by Marius should be continued in naming features on the satellites. Features would be named for persons and places associated with the classical myths of the persons for whom the satellites are named. If additional names were needed, the Task Group planned to use names from mythologies of ethnic groups, chosen according to their terrestrial latitudinal location: Tropical ethnologies would be used for Io, names from European myths for Europa, names from other temperate zone mythologies for Ganymede, and names from mythologies of polar ethnic groups for Callisto. However, when Voyager arrived at Jupiter, it imaged strange new worlds of fire and ice that were unlike any seen in the inner solar

Figure 4.4 (facing page) A. Nomenclature as shown on a shaded relief rendition of Venus; nomenclature also shown on Topographic and Shaded Relief Map of Venus *(U.S. Geol. Survey Misc. Inv. Ser. Map I-1562). B. Venusian nomenclature on the* Magellan Planning *chart (USGS, I-2041, sheet 1) shows names chosen from Pioneer Venus imagery, as well as many new features discriminated by the Soviet radar instruments on the Venera 15–16 spacecraft.*

system. Obviously, the plan for naming features on these remarkable bodies would have to be modified, and new terms chosen to define adequately the features Voyager was imaging.

Evidence of active volcanism seen on the images of Io's caldera-pocked, sulfurous surface suggested that a volcanic theme be adopted for that satellite. Paterae (calderas) and nine documented zones of active volcanism were named for fire, sun, and blacksmith ("fire worker") gods and heroes of worldwide ethnic groups. For example, an active eruptive vent was named Pele, after the Hawaiian goddess of the volcano, and a patera (caldera) was named for the Norse trickster character who gave fire to humankind: Loki Patera. Other features were named for people and places associated with the Io myth (Figure 4.5).

Resolution is poor in photographs of Europa, and names associated with the Europa myth sufficed to identify the relatively few features that were clearly enough defined to require a name (Figure 4.6). However, three new feature terms were adopted for strange features seen on the satellite: *flexus* for cuspate linear features, *linea* for straight features, and *macula* for circular dark areas (Table 4.1).

On the other hand, Ganymede has many prominent features. Although the Task Group started to name features from the Ganymede myth, the list of available names was soon exhausted. A new theme, based on the names of persons and places in the mythology of ancient Near Eastern civilizations, was selected to name these features. For example, the longitude-determining small crater Anat is named for the Assyro-Babylonian goddess of dew, and a prominent system of faults is named Lakmu Fossae for the Babylonian dragon of chaos (Figure 4.7). Another new term, *facula* (Table 4.1), was adopted for bright circular features, and a term used originally in the early Martian nomenclature, *regio,* was used for large, nebulous dark regions. The faculae are named for ancient Near Eastern places (Punt Facula is named for a land east of ancient Egypt). The discoverers of Jovian satellites are honored in the naming of the regiones; for example, Perrine Regio honors the discoverer of Himalia and Elara, two small Jovian satellites.

The surface of the outermost of the Galilean satellites, Callisto, is the only one that superficially resembles the more familiar inner planets; its heavily cratered surface is somewhat reminiscent of the Moon, Mercury, and Mars, though the large multiringed structures seen on Callisto are quite different from ringed features on any other planets. Craters on Callisto are named for mythological deities, heroes, and places in the rich Scandic epics, or from Alaskan and Patagonian mythologies. Ringed features, such as Valhalla, are named for mythical Norse or Alaskan abodes of the dead, and one catena, Gipul Catena, is named for a Norse river (Figure 4.8).

The innermost Jovian satellite, Amalthea, is small and so irregularly shaped that a coordinate system has not been determined for it; four craters shown in a diagram of its surface are named for characters in the Greek myth about Amalthea (Figure 4.9).

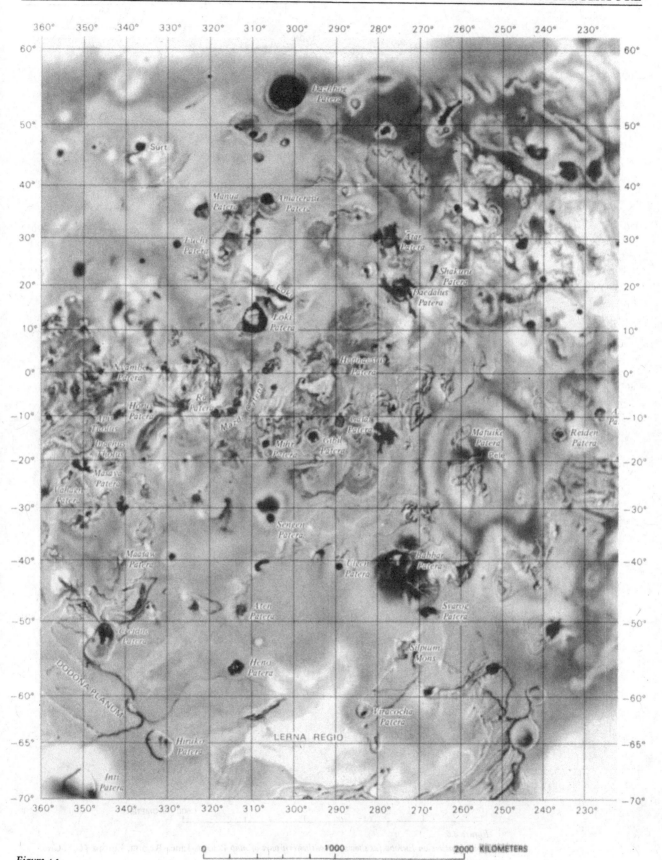

Figure 4.5
Nomenclature on Io, as shown in part of Preliminary Pictorial Map of Io *(U.S. Geol. Survey Misc. Inv. Map I-1240).*

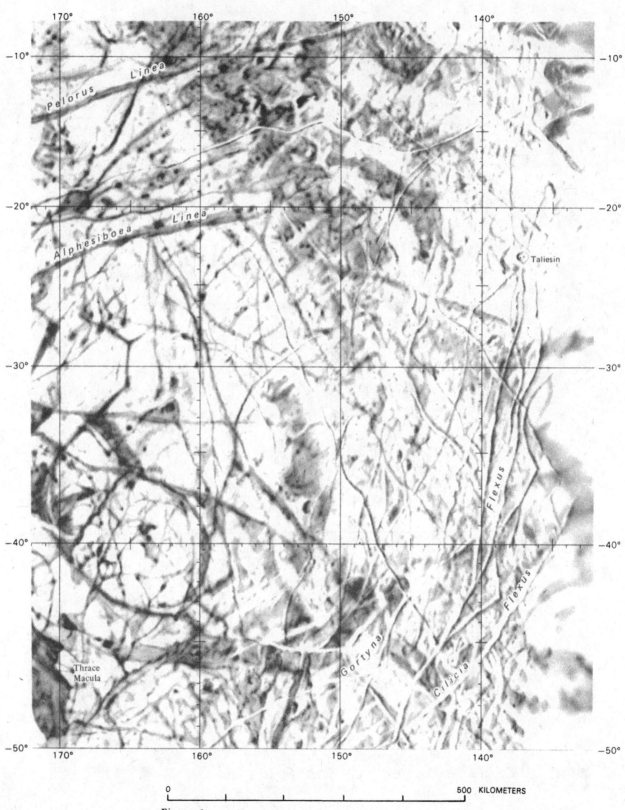

Figure 4.6
Nomenclature on Europa, as shown in southestern part of map Pelorus Linea Region, Europa *(U.S. Geol. Survey Misc. Inv. Ser. Map I-1493).*

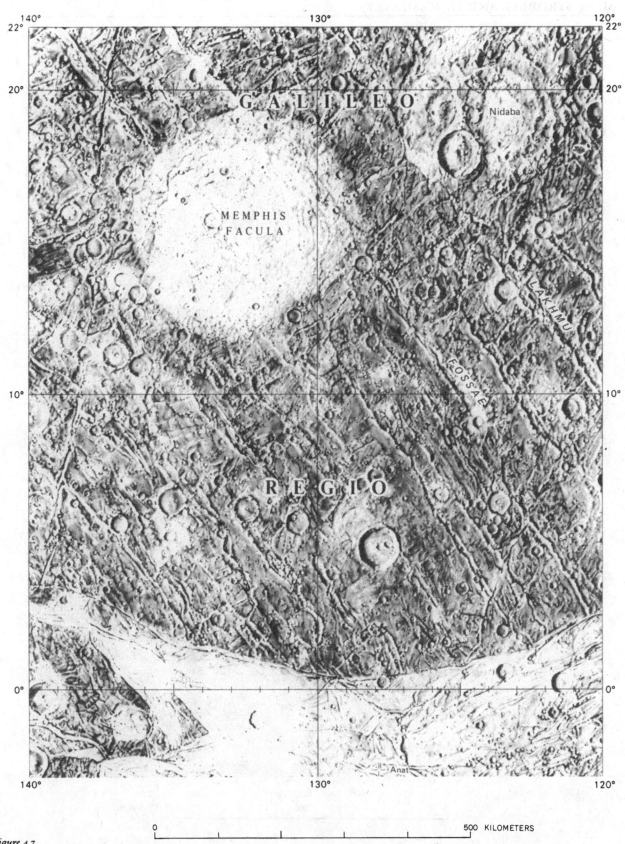

Figure 4.7
Nomenclature on Ganymede, as shown in northwestern part of Memphis Facula quadrangle, Ganymede
(U.S. Geol. Survey Misc. Inv. Ser. Map I-1536).

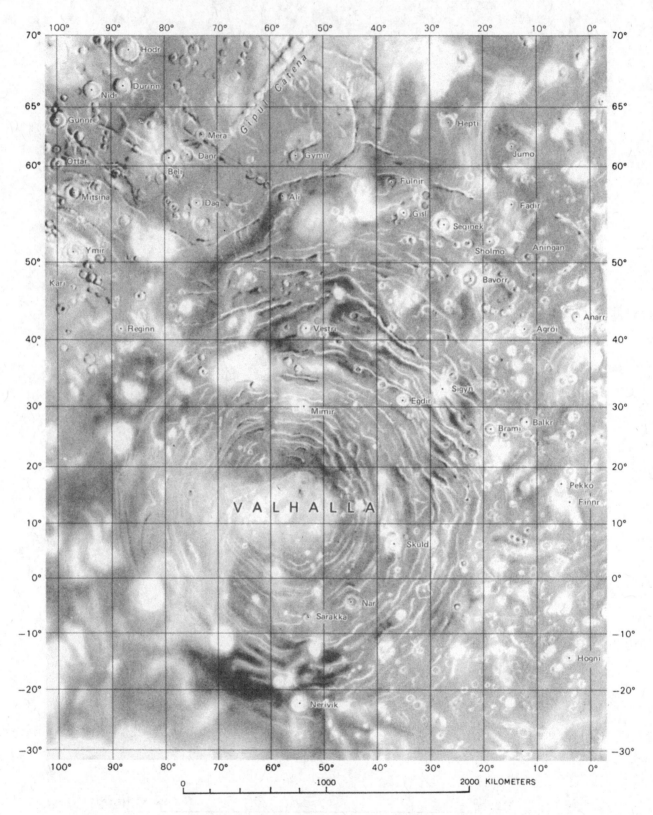

Figure 4.8
Nomenclature on Callisto, as shown in north-central part of Preliminary Pictorial Map of Callisto *(U.S. Geol. Survey Misc. Inv. Ser. Map I-1239).*

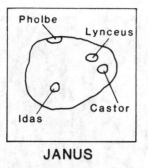

JANUS

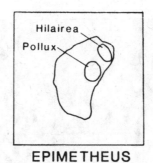

EPIMETHEUS

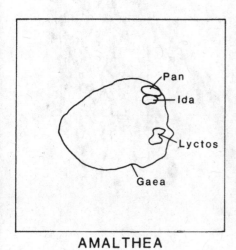

AMALTHEA

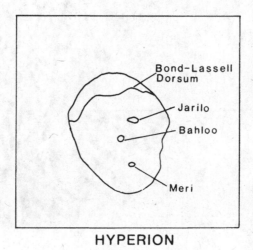

HYPERION

Figure 4.9
Nomenclature on the small Jovian satellite Amalthea, and the small Saturnian satellites Janus, Epimetheus, and Hyperion.

4.9.   SATURNIAN SATELLITES: 1980–2

The mythological theme used to name the Jovian satellites was also used to name the Saturnian satellites. Huygens was the first to discover a Saturnian satellite, in 1655; he named it Titan. The Italian-born French astronomer Giovanni Cassini discovered and named Iapetus, Rhea, Dione, and Tethys in the 1670s and 1680s. William Herschel discovered Mimas and Enceladus in 1789, the American astronomers George and William Bond discovered Hyperion in 1848, and the American astronomer William Pickering discovered Phoebe in 1898. Finally, the French astronomer Auduoin Dollfus discovered, in 1966, what has turned out to be one of two satellites sharing very similar orbits. Dollfus chose the name Janus for the satellite he observed. Its "twin" satellite was discovered and named Epimetheus by the Voyager team in 1981. At this time the IAU received many suggestions to change Janus' name to Prometheus; however, the name chosen by its discoverer prevailed, and the name Prometheus was assigned to another small satellite.

When choosing names for features on the small Saturnian satellites, the convention of honoring all humankind was continued. Janus and Epimetheus

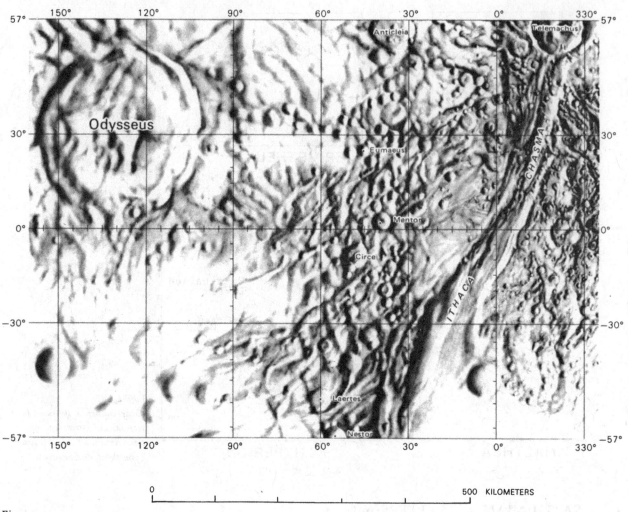

Figure 4.10
Nomenclature on Tethys, as shown in western equatorial section of Preliminary Pictorial Map of Tethys *(U.S. Geol. Survey Misc. Inv. Ser. Map I-1487).*

share similar orbits and so were given names from the myth about Castor and Pollux, the famous Greek/Roman mythological twins (Figure 4.9). Features on Hyperion, the father of the Sun in Greek mythology, are named for Sun or Moon gods of other ethnic groups. (Figure 4.9).

On the larger Saturnian satellites features are also named from world-wide mythologies, but the scheme for applying the convention is different from the Jovian system. On Saturn epic stories from various cultures were selected as the theme for naming features; each satellite was assigned an epic, and names of persons (craters) and places (other features) from the epic are assigned to individual features. Thus, Odysseus and Ithaca Chasma, names from Homer's *Odyssey* (Bates, 1929), are the most prominent features on Tethys (Figure 4.10); Aeneas and Carthage Linea, from Virgil's *Aeneid* (Mandelbaum, 1964), are prominent on Dione (Figure 4.11); Shahrazad and Samarkand Sulci, from Burton's (1900) compilation and translation of the *Thousand Nights and One Night,* are found on Ence-

122

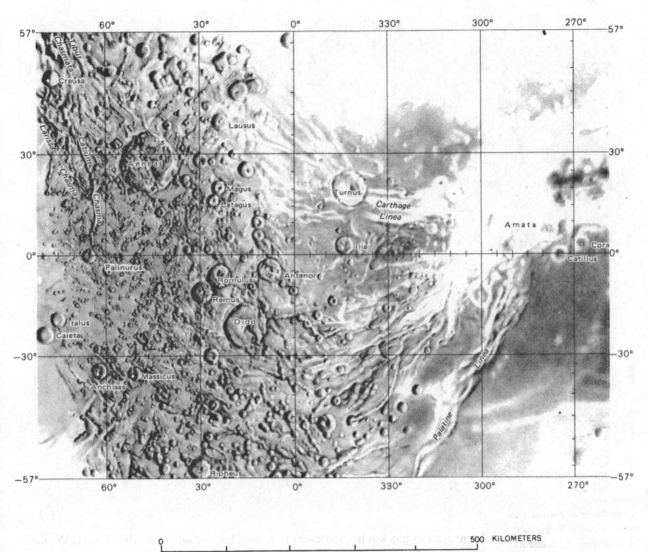

0 _____ 500 KILOMETERS

Figure 4.11
Nomenclature on Dione, as shown on central part of Preliminary Pictorial Map of Dione *(U.S. Geol. Survey Misc. Inv. Ser. Map I-1488).*

ladus (Figure 4.12); and Charlemagne and Roncevaux Terra, from Sayers' translation (1967) of the medieval French epic, *Song of Roland,* are featured on Iapetus (Figure 4.13).

An attempt to find Asiatic, South American, or African epics that were not related to living religions proved unsuccessful. In order to internationalize the Saturnian satellite nomenclature, features on Rhea were named from creation stories of Asian, African, and South American ethnic groups (Figure 4.14). Here the largest crater is named for the Japanese god Izanagi.

The convention of naming a prominent feature for the satellite's discoverer, developed for the Jovian satellites, was also continued on three of the Saturnian satellites. On Mimas, where other features are named from Baines' (1962) translation of the medieval English and French epic *Le Morte D'Arthur,* a very

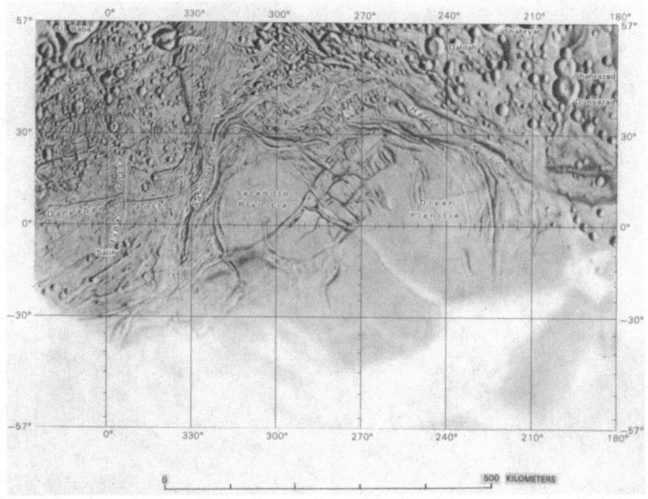

Figure 4.12
Nomenclature on Enceladus, as
shown on part of Preliminary
Pictorial Map of Enceladus
(U.S. Geol. Survey Misc. Inv.
Ser. Map I-1485).

large crater (300 km in diameter) is named for the satellite's discoverer, William Herschel (Figure 4.15). A dark area, which covers half the equatorial region of Iapetus, extends into the lower left corner of Figure 4.13; it is named Cassini Regio, for Iapetus' discoverer. On Hyperion a ridge is named for the American astronomers George and William Bond and the amateur British astronomer William Lassell; these men discovered the small satellite on the same night, working independently.

Elements of Saturn's ring system also are named. In 1655 Huygens was able to observe Saturn using a telescope with sufficient power to discriminate a disk-shaped ring structure around the planet. (Galileo in 1610 had thought there were two satellites of Saturn; other astronomers in the intervening period had advanced other theories to explain the peculiar object or objects observed through primitive telescopes.) In 1675 Cassini recognized a division in the disk or ring, and the main ring structure was known thereafter as the A (outer) and B (inner) rings. Several astronomers observed divisions in the A ring but could not agree on their exact position (or positions) until 1837, when the German astronomer Johann Franz

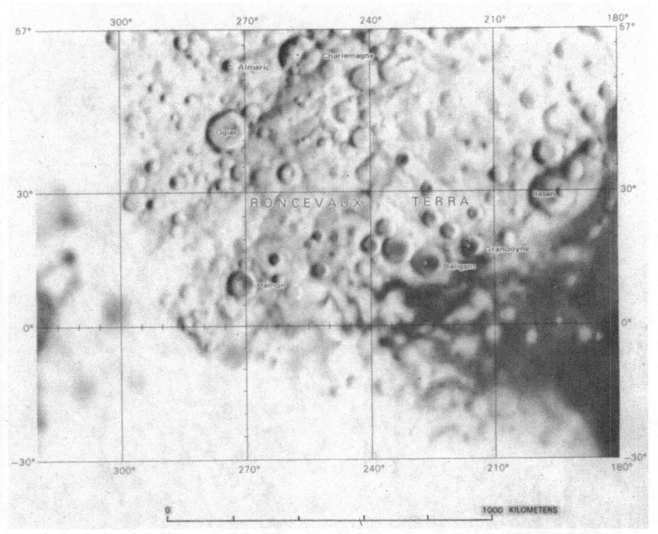

Figure 4.13
Nomenclature on Iapetus, as shown in eastern equatorial region of Preliminary Pictorial Map of Iapetus *(U.S. Geol. Survey Misc. Inv. Ser. Map I-1486).*

Encke observed one division that appeared to be continuous through the ring. In November 1850 a third ring, located between the B ring and the planet, was officially "discovered" by the Bonds and their assistant C. W. Tuttle, and the British astronomer William Dawes. These scientists had recognized elements of the third ring but had not officially designated the elements as constituting a distinct ring feature. In the following month Lassell named this ring the Crepe, or C, ring. In 1859 the British astronomer James Clerk Maxwell hypothesized that the rings comprised a "multitude of small, disconnected satellites," rather than being either solid or liquid in form. By combining a telescope with a spectroscope and camera, the American astronomer James E. Keeler, in 1895, proved Maxwell's hypothesis when he obtained pictures that provided spectroscopic proof of the rotation of the Saturnian rings. The three rings have been called the A, B, and C (Crepe) rings since they were recognized. In 1982 the IAU named two divisions in the ring system for Cassini and Encke and three

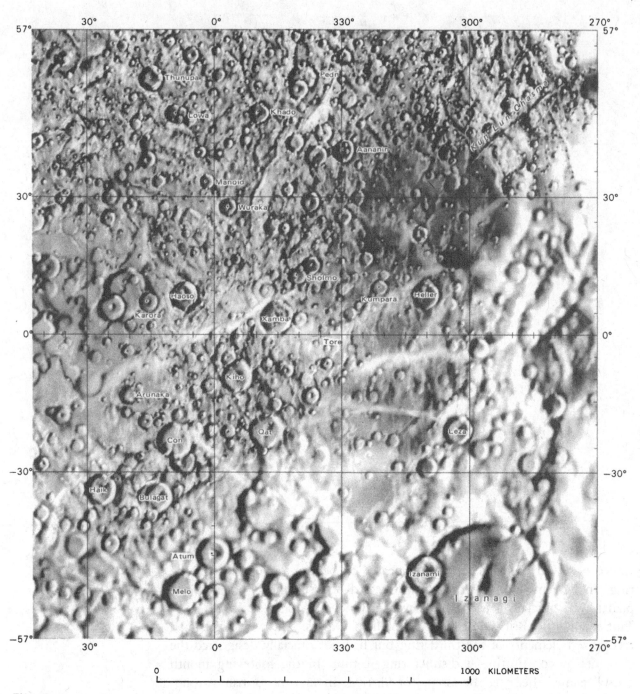

Figure 4.14
Nomenclature on Rhea, as shown in northwestern equatorial region of Preliminary Pictorial Map of Rhea *(U.S. Geol. Survey Misc. Inv. Ser. Map I-1484).*

ring gaps for Huygens, Maxwell, and Keeler, to honor the scientists who first discovered ring features. In 1988 the IAU adopted the name Coombo for a gap in the C ring, to honor Giuseppe Colombo, who devised the trajectory for the multiple flybys of Venus–Mercury for the Mariner 10 mission (Figure 4.16).

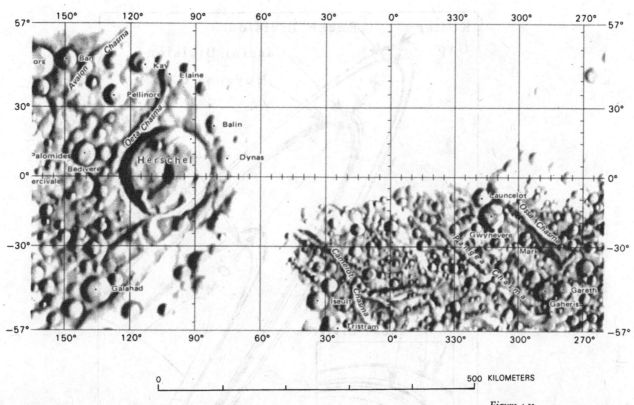

Figure 4.15
Nomenclature on Mimas, as
shown on eastern part of Preliminary Pictorial Map of Mimas (U.S. Geol. Survey Misc.
Inv. Ser. Map I-1489).

4.10. EXPANSION OF EXISTING NOMENCLATURE: 1981–9

Although no active missions to a previously unexplored planet were
launched during the period between the two Voyager encounters with Saturn in November 1980 and August 1981 and the arrival of Voyager 2 at
Uranus in January 1986, the nomenclature Task Groups and Working
Group were very active. Besides the names chosen and adopted for features
on the Jovian satellite features, discussed above, names were approved for
several lunar features that had been shown on lunar maps but were never
officially adopted by the IAU. A *Catalogue of Lunar Nomenclature* was also
compiled by Andersson and Whitaker (1982). Thirty-seven names were
adopted for Mercurian craters to meet a rejuvenated interest in the geology
of Mercury. A very successful Soviet mission to Venus produced higher-
resolution images of 25 percent of the planet (including the previously
unobserved north polar region). Two new feature terms, *corona* for ovoid
structures and *tessera* for mosaic-like terrain, were adopted and more than
four hundred additional features were named from the Soviet radar images.
Feature names adopted in 1985 and 1988 are shown, with features named
earlier, on the Magellan Planning Chart (Figure 4.4B) published in 1989
by the USGS. Two new series (at 1:2,000,000 and 1:500,000 scales) of

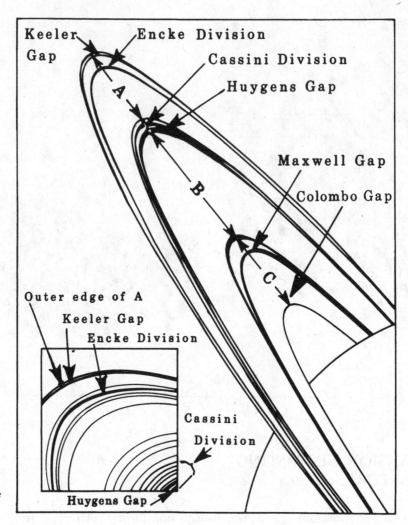

Figure 4.16
Nomenclature of elements in the
Saturnian ring system.

Martian maps discriminated many interesting features for which names were requested by investigators. Higher-resolution maps of Ganymede, Callisto, and Io were also produced during this period, and geologic analyses of these regions are in progress; many new names have been requested by scientists engaged in these studies.

A new Task Group for Small-Body, Surface-Feature Nomenclature was appointed in 1985 at the New Delhi meeting of the IAU; in the future this group will propose names for surface features on comets and asteroids. The names of individual asteroids are chosen by their discoverers; comets are named for their discoverer or discoverers. The naming process for these small bodies is overseen by another arm of the IAU, the Minor Planets Names Committee (MPNC); this group is part of Commission 20 and is chaired by Brian G. Marsden (Smithsonian Astrophysical Observatory, Cambridge, Massachusetts). Rules for naming these small bodies are much less stringent than those governing planetary and satellite feature names;

the discoverer is allowed wide latitude as long as the name does not offend any other person or group. Duplication between names chosen for comets and asteroids and those chosen for features on planets and satellites is allowed because both sets are large and their themes are similar; duplication is unavoidable.

Another nomenclature group, the Satellite Nomenclature Liaison Committee (SNLC), was formed at the Patras meeting of the IAU to work with Commission 20, the MPNC, and the WGPSN in naming newly discovered satellites. Several years earlier, a satellite was discovered and named before its orbit had been established unequivocally. In the early 1970s, duplicate "discoveries" were sometimes made. In order to avoid such problems in the future, in 1979 Commission 20 asked the MPNC to assign to each newly observed body a temporary designation that indicates the year and order of its discovery; for example, the first unnamed new satellite of Uranus discovered by the Voyager crew is designated 1985 U1, the second is 1986 U1, the third is 1986 U2, etc. After Commission 20 has verified the orbit of a new body, SNLC assigns the satellite a permanent designation composed of a capital letter that designates the parent planet and a Roman numeral that indicates either the order of discovery or the distance from the planet: Amalthea is JV, Mimas is SI, and Janus is SX. After orbit verification, the SNLC, working with the discoverer and the WGPSN, proposes a name for the newly discovered satellite. For example, 1985 U1 is provisionally named Puck (Table 4.4).

4.11. URANIAN SATELLITES: 1984–8

During the period between missions, the Outer Planets Task Group also prepared preliminary lists of names to be used when Voyager arrived at Uranus. The five known satellites of Uranus had been named for characters in Shakespeare or Pope; four were the names of fairies. Thus, lists of names were compiled for Shakespearean and spirit categories.

The Uranus encounter proved to be as exciting and interesting as were previous stages of the Voyager mission. Late in 1985 photos were obtained of an irregularly shaped, small (85 km in long dimension) satellite that was informally dubbed Puck by the Voyager team members. The name seemed to suit the satellite (Figure 4.17), and the IAU committees gave their approval in 1988. Puck is not only the name of a mischievous spirit in Shakespeare's *A Midsummer Night's Dream* but also the name of a class of mischievous spirits in several northern European mythologies, so three relatively prominent craters on Puck are named for his counterparts in other ethnic groups (Figure 4.17).

Nine other small satellites, mere specks of light on Voyager images, are named for Shakespearean heroines, even though more than 8000 letters were received by IAU members asking that these small satellites be named for deceased members of the crew on the American space vehicle *Challenger,*

Table 4.4. *Satellite names and designations.*

Planet	Temporary designation	Permanent designation	Satellite name
Earth			Moon
Mars		MI	Phobos
		MII	Deimos
Jupiter		JI	Io
		JII	Europa
		JIII	Ganymede
		JIV	Callisto
		JV	Amalthea
		JVI	Himalia
		JVII	Elara
		JVIII	Pasiphae
		JIX	Sinope
		JX	Lysithea
		JXI	Carme
		JXII	Ananke
		JXIII	Leda
	1979 J2	JXIV	Thebe
	1979 J1	JXV	Adrastea
	1979 J3	JXVI	Metis
Saturn		SI	Mimas
		SII	Enceladus
		SIII	Tethys
		SIV	Dione
		SV	Rhea
		SVI	Titan
		SVII	Hyperion
		SVIII	Iapetus
		SIX	Phoebe
	1980 S1	SX	Janus
	1980 S3	SXI	Epimetheus
	1980 S6	SXII	Helene
	1980 S13	SXIII	Telesto
	1980 S25	SXIV	Calypso
	1980 S28	SXV	Atlas
	1980 S27	SXVI	Prometheus
	1980 S26	SXVII	Pandora
Uranus		UI	Ariel
		UII	Umbriel
		UIII	Titania
		UIV	Oberon
		UV	Miranda
	1986 U7	UVI	Cordelia
	1986 U2	UVII	Juliet
	1986 U9	UVIII	Bianca
	1986 U3	UIX	Cressida
	1986 U6	UX	Desdemona
	1986 U8	UXI	Ophelia
	1986 U1	UXII	Portia
	1986 U4	UXIII	Rosalind
	1986 U5	UXIV	Belinda
	1985 U1	UXV	Puck
Neptune	1989	NI	Triton
	1989	NII	Nereid
Pluto	PI		Charon

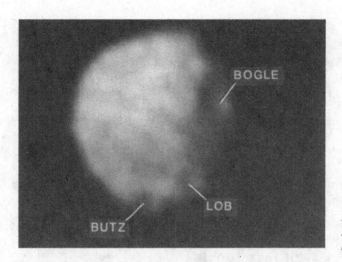

Figure 4.17
Nomenclature on Uranian satellite "Puck."

which blew up a few days after Voyager's closest approach to Uranus. Members of the IAU committees preferred to stay with the Shakespearean tradition; instead, they have assigned the names of *Challenger* crew members to seven craters on the Moon's farside, within crater Apollo and near craters named for other astronauts.

Although William Herschel discovered Uranus (in 1781) and two of its satellites, Titania and Oberon (in 1783), he was not responsible for the present names of either the planet or its satellites. He had chosen the name Georgium Sidus (Georgian planet) for his discovery, to honor the king of England. However, the name did not stick, and the German astronomer Johann Bode chose a more conventional name, Uranus, for the planet, in the same year that it was discovered. Uranus is a Roman god, the father of Saturn, who in turn is the father of Jupiter. So Bode's choice of name simply continued the "family tree" backward in mythological time, as the planets stretch outward from the sun. Two additional satellites were discovered by William Lassell in 1851. The four satellites were named in 1851 by Lassell, who in turn accepted names suggested to him by Herschel's son, Sir John Herschel: Ariel, Umbriel, Titania, and Oberon. A fifth satellite was discovered in 1948 by the American astronomer Gerard Kuiper; he named the satellite Miranda, for the human heroine of Shakespeare's comedy *The Tempest*, thus continuing the Shakespearean theme started by John Herschel. The Shakespearean/spirit categories were used to name features on Miranda, as well as on the other satellites. Craters on Miranda are named for other humans in *The Tempest*, and other features were named for places prominent in Shakespearean plays (Figure 4.18). Ariel is a "good" or bright spirit in both *The Tempest* and Alexander Pope's *Rape of the Lock*. Features on Ariel, the Uranian satellite with the brightest albedo value (Figure 4.19), are named for good or bright spirits; craters are named for individual spirits, and the valles and chasmata are named for classes of light spirits. Umbriel's albedo is very dark, and the character in Pope's mock-

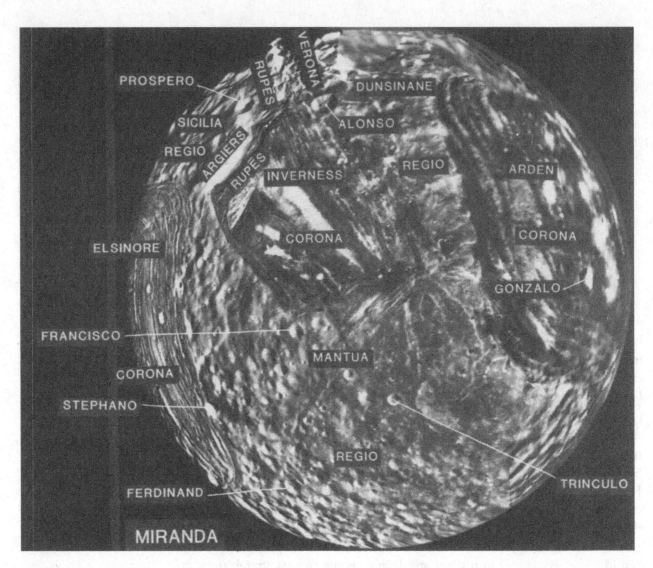

Figure 4.18
Nomenclature on Uranian satellite Miranda. Satellite diameter averages approximately 562 km (Davies et al., in press); south pole is in center of image.

epic poem is the personification of dark or evil thoughts (or spirits). Craters on Umbriel are named for evil spirits or those who dwell underground (Figure 4.20). Craters on Titania, the fairy queen of Shakespeare's *A Midsummer Night's Dream*, are named for female Shakespearean characters, mostly of minor importance; other features are named for places prominent in Shakespearean plays (Figure 4.21). Finally, on Oberon, named after the king of the night in *A Midsummer Night's Dream*, craters are named for the famous Shakespearean tragic heroes (Figure 4.22).

Uranian rings discovered from Earth-based observations are currently designated by both Greek letters and Roman numerals. Rings and ring-arcs discovered by Voyager were given temporary designations such as 1986 UR–1. Although much discussion has been held regarding a possible method of

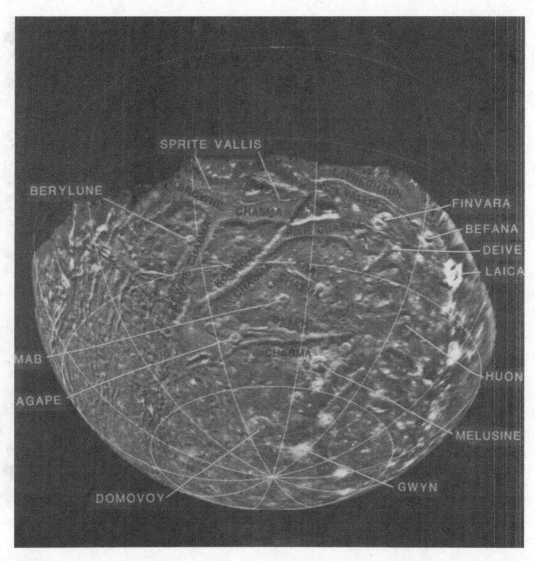

Figure 4.19
Nomenclature on Uranian satel-
lite Ariel. Satellite diameter is
1158 km (Davis et al., 1987);
image of area near south pole.
See also USGS Map I-1920.

renaming the rings, no proposal has received strong support. Figure 4.23 shows the present informal nomenclature.

4.12. NEPTUNE'S SATELLITES: 1989

While Voyager 2 continued onward from Uranus to its planned encounter with Neptune in August 1989, the Outer Solar System Task Group prepared a tentative list of names for features that might be disclosed on the surface of Neptune's largest moon, Triton. It was also possible that additional small satellites might be discovered, so a list of appropriate names was also selected against this eventuality. The lists will be used; once again, Voyager has disclosed heretofore unimagined features on Triton's cold surface. In addition, several new satellites and tenuous rings were also discovered.

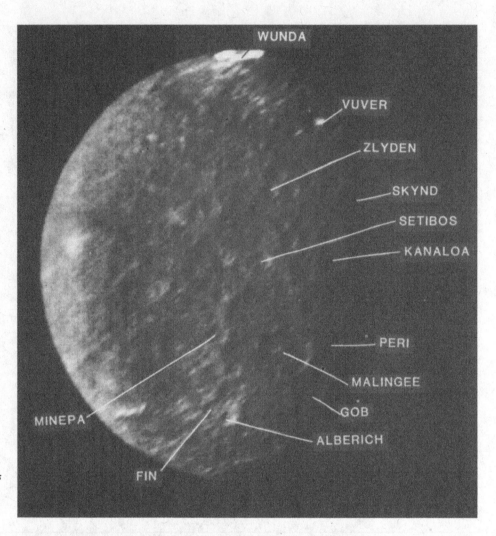

Figure 4.20
Nomenclature on Uranian satel-
lite Umbriel. Satellite diameter is
1172 km (Davies et al., 1987);
image of area near south pole.
See also USGS Map I-1920.

In expanding the nomenclature of any planet or satellite, consideration is always given to the established nomenclatural theme. The name bank for Neptune and its satellites was aquatic, so the new nomenclature is also aquatic. A list of names of classical sea gods and spirits, like Triton and Nereid, has been compiled to help the discoverers name the new satellites; however, the final choice of a name rests with the discoverer. A list of aquatic persons, places, or things (chosen from worldwide ethnic groups) has been compiled as possible names for Triton's strange features.

4.13. FUTURE PLANS

The need for names will continue as long as maps are prepared at higher resolution or new images are obtained of planetary or satellite surfaces. New

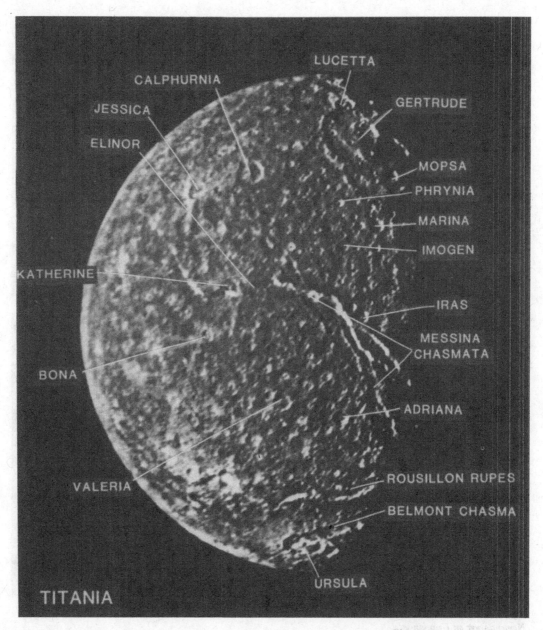

Figure 4.21
Nomenclature on Uranian satellite Titania. Satellite diameter is 1580 km (Davies et al., 1987); image of area near south pole. See also USGS Map I-1920.

enhancements of Viking high-resolution images and maps made from them allow identification of many additional features for which planetary scientists will need names. If the planned Observer and Lander/Rover/Returned Sample missions to Mars become a reality, our knowledge of that planet will take the same quantum leap that occurred when Apollo astronauts studied and then landed on the Moon. The Galileo missions will acquire images of Jupiter and its satellites that will surpass those obtained by Voyager as much as Viking's pictures surpass those of Mariner 9; new names for planetary atmospheric features as well as satellite features will be needed. Improved resolution will

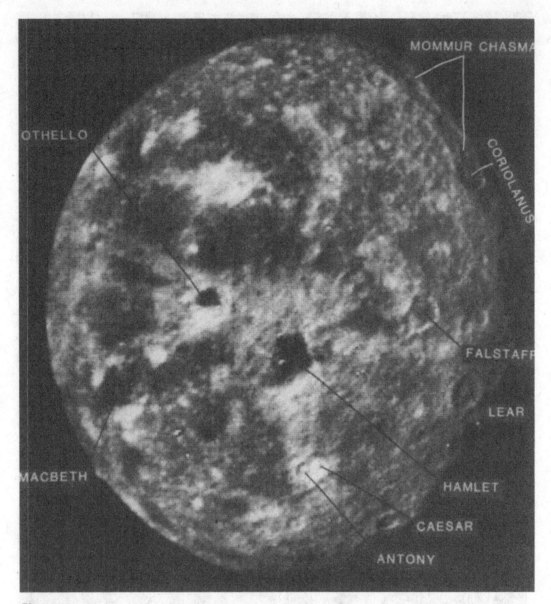

Figure 4.22
Nomenclature on Uranian satel-
lite Oberon. Satellite diameter is
1524 km (Davies et al., 1987);
image of area near south pole.
See also USGS Map I-1920.

permit the Magellan radar to discriminate new features on the Venusian surface
and also obtain stereoscopic images that will provide a better understanding
of all features.

4.14. SUMMARY AND CONCLUSIONS

Naming new features is a time-consuming job that must be done carefully.
In the past, planetary astronomers and other planetary scientists were usually
the only people who requested and used planetary names. However, space

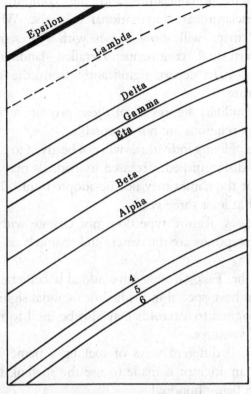

URANIAN RING SYSTEM

Figure 4.23
Present informal nomenclature on Uranian rings.

exploration has become common and intensely competitive in the latter part of the twentieth century. A much larger number of scientists from diverse disciplines, as well as a large coterie of laymen from the world community, are now intensely interested in all aspects of space study, including the nomenclature applied to now familiar planetary features.

Input from the public is welcomed by the IAU committee members. However, individuals wishing to submit names are encouraged to familiarize themselves with the following rules and constraints adopted by the IAU:

1. Names should be simple, clear, unambiguous, and easy to pronounce and spell.
2. The number of names should be kept to a minimum and governed by the needs of the scientific community.
3. Duplication of a name on more than one body should be avoided.
4. In general, individual names should be single words expressed in the language of origin.
5. Where possible, consideration is given to the traditional aspects of nomenclature, provided this does not cause confusion.

6. Solar system nomenclature and the committees that govern the nomenclature are international in scope. Wherever possible, individual maps will show names with representation from the various terrestrial continents. Detailed landing site maps of small areas may depart significantly from the general rules of nomenclature.

7. Names of military figures or of those prominent in any of the six main living religions are not acceptable.

8. The names of living individuals will not be given to planetary features. Names of distinguished, deceased individuals may be proposed for craters, but the names may not be adopted until the individual has been dead at least three years.

9. In most cases, feature type does not change with the size of the feature. Exceptions are the craters and channels on Mars and craters on Venus.

10. Recently the Task Groups have added a constraint against using names that have special individual or national significance.

11. Names assigned to asteroids may also be used to name features on planets or satellites.

12. When several different ways of spelling a name are found in the literature, an attempt is made to use the spelling that was used by the person being honored.

13. Wherever possible, the most authoritative translation or compilation of an epic is used in deciding which spelling of a name to use.

The above rules apply to all names chosen for features on planets and satellites. They summarize nearly three hundred years of experience in planetary nomenclature and are based on past, sometimes bitterly contentious, experience. Occasionally a rule has been broken, but only with the concurrence of the full General Assembly of the IAU, when members of the appropriate Task Group and the Working Group agree that existing circumstances are so extraordinary that the general rule should be broken. These rules have worked well in the past. Planetary nomenclature is an effective tool at present because the choice of new names is done in a fair and evenhanded way by an international group of experts. It will continue to be useful as long as the group remains international in scope and adheres to the rules and constraints developed over the past three centuries.

4.15. REFERENCES

Andersson, Leif, and Whitaker, E. A. 1982. *NASA Catalogue of Lunar Nomenclature*. National Aeronautics and Space Agency Reference Publication 1097.

Antoniadi, E. M. 1930. *La planète Mars*. Paris: Librairie Scientifique Hermann.

Antoniadi, E. M. 1934. *La planète Mercure*. Paris: Gauthier-Villars, Fig. 5, p. 26.

(Translated by Patrick Moore. 1974. Shaldon Devon: Keith Reid.)

Arthur, D. W. G., Agnieray, A. P., Horvath, R. A., et al. 1963. The system of lunar craters, quadrant I. *Communications of the Lunar and Planetary Laboratory* (30): 71–8, 4 unnumbered appendixes, 12 unnumbered maps.

Arthur, D. W. G., Agnieray, A. P., Horvath, R. A., et al. 1964, The system of lunar craters, quadrant II. *Communications of the Lunar and Planetary Laboratory* 3 (40): 1–59, 12 unnumbered maps.

Arthur, D. W. G., Agnieray, A. P., Pellicori, R. H., et al. 1965. The system of lunar craters, quadrant III. *Communications of the Lunar and Planetary Laboratory* 3 (50): 61–2, catalog p. 1–146, 12 unnumbered maps.

Arthur, D. W. G., Pellicori, R. H., and Wood, C. A. 1966. The system of lunar craters, quadrant IV. *Communications of the Lunar and Planetary Laboratory* 5 (70): 1, catalog p. 1–208, 12 unnumbered maps.

Baines, Keith. 1962. *Malory's Le Morte d'Arthur*. New York: Mentor.

Bates, Herbert, transl. 1929. *The Odyssey of Homer*. New York: Harper Brothers.

Blagg, Mary. 1913. *Lunar Formations Named or Lettered in the Maps of Neison, Schmidt, and Mädler*. Edinburgh: Neill.

Blagg, Mary, and Müller, Karl. 1935. *Named Lunar Formations*. London: Percy, Lund, Humphries.

Blunck, Jürgen. 1977. *Mars and Its Satellites*. Hicksville, N.Y.: Exposition Press.

Blunck, Jürgen. 1982. *Mars and Its Satellites*, 2nd ed. Hicksville, N.Y.: Exposition Press.

Burton, Richard. 1900. *Alf Laylah Wa Laylah; The Book of a Thousand Nights and a Night*. 16 vols. Denver: Harper.

Davies, M. E., Colvin, T. R., Katayama, F. Y., 1987. The control networks of the satellites of Uranus. *Icarus*, v. 71, p. 137–147.

IAU. 1960. Commission 16, Physical study of the Planets. In 10th General Assembly, Moscow, 1958. *Transactions: International Astronomical Union Proceedings* 10: pl. I, II.

IAU. 1971. Commission 16, Physical study of the planets and satellites. In 14th General Assembly, Brighton, 1970. *Transactions: International Astronomical Union Proceedings* 14: 128–9.

IAU. 1977. Working Group for Planetary System Nomenclature. In 16th General Assembly, Grenoble, 1976. *Transactions: International Astronomical Union Proceedings* 16B: 363–9.

IAU. 1986. Working Group for Planetary System Nomenclature. In 19th General Assembly, Delhi, 1985. *Transactions: International Astronomical Union Proceedings* 19B: 339–53.

IAU. 1988. Working Group for Planetary System Nomenclature. In Reports on Astronomy. *Transactions: International Astronomical Union Proceedings*, pp. 703–8.

IAU. In press. Working Group for Planetary System Nomenclature. In 20th General Assembly, Baltimore, 1988. *Transactions: International Astronomical Union Proceedings* 20B.

Jet Propulsion Laboratory. 1971. *MM '71 Mars Planning Charts, North Polar Region, Equatorial Region, and South Polar Region*.

Mandelbaum, Allen, transl. 1962. *The Aeneid of Virgil*. 2 vols. New York: Macmillan.

Menzel, D. H., Minnaert, M., Levin, B., and Dollfus, A. 1971. Report on lunar nomenclature. *Space Science Reviews* 12 (2): 136–86.

Murray, J. B., Dollfus, A., and Smith, B. 1972. Cartography of the surface markings of Mercury. *Icarus* 17 (3): 581, Fig. 4.

Sayers, D. L., transl. 1967. *Song of Roland*. Harmondsworth: Penguin.

Schiaparelli, G. V. 1877–8. Osservazioni astronomiche e fisiche sull'asse di rotazione e sulla topografia del planeta Marte. In *Atti della R. Academia dei Lincei, Memoria della cl. di scienze fisiche*, Memorie 1, ser. 3, v. 2, pp. 308–439.

Schiaparelli, G. V. 1880–1. Osservazioni astronomiche e fisiche sull'asse di rotazione e sulla topografia del planeta Marte. In *Atti della R. Academia dei Lincei, Memoria della cl. di scienze fisiche*, Memorie 2, ser. 3, v. 10, pp. 281–387.

United States Geological Survey. 1989. Topographic Map of Part of the Northern Hemisphere of Venus, Sheet 1 of 3. Scale, 1:15,000,000.

Whitaker, Ewen A. 1985. *The University of*

Arizona's Lunar and Planetary Laboratory – Its Founding and Early Years. Tucson: University of Arizona Printing Reproductions Dept.

Whitaker, E. A., Kuiper, G. P., Hartmann, W. K., and Spradley, H. L. 1963. *Rectified Lunar Atlas*. Tucson: University of Arizona Press.

Working Group for Planetary System Nomenclature. 1986. *Annual Gazetteer of Planetary Nomenclature*. U.S. Geological Survey Open-File Report 84–692.

5

Geodetic control

MERTON E. DAVIES

5.1. BACKGROUND

The science of geodesy is concerned with the determination by observation
and measurements of the exact positions of points over large areas of the
Earth's surface. Geodesy also deals with the size and shape of the Earth and
variations in gravity. Like geology, the science of geodesy has in recent years
been extended to encompass other bodies of the solar system. On Earth,
geodetic or control networks are established by land surveys, aerial or space
photography, and more recently, by radio Doppler measurement of signals
from artificial satellites whose orbits are precisely determined. The control
networks are used for positional data in the production of maps.

On Earth, the surveyor's brass control point marker is a familiar sight;
however, surveying has not yet been found practical on other planets. For
planetary work, the photogrammetric techniques developed for extending
control using aerial photography are employed. The control points identified
and measured on aerial photographs might be a church steeple, a road intersection, or any conspicuous landmark. On planets and satellites, the most
common control points are the centers of craters as defined by their rims.
Analytical photogrammetry is then used to compute the latitude and longitude
of the control points. This computation sometimes involves blocks of several
hundred photographs and thousands of points.

This section will discuss the development of geodetic control on the Moon
and Mars based on telescopic data. After a historical review of developments
before the introduction of photography, modern techniques using photography will be discussed. The final topic of this section is the status of control
on Mars before the exploration by spacecraft. Additional sections discuss the
coordinate systems of the planets and satellites, analytical triangulation, and
finally, the control networks of these bodies.

5.1.1. *Early lunar control*

In 1610 Galileo Galilei (1564–1642) published the first drawings of the
Moon, based on observations made with the aid of a telescope. In the years

that followed, many beautiful drawings and maps appeared. However, in the modern sense these did not qualify as maps: They did not have a reference coordinate system, and positional data were not based on measurements. It was Tobias Mayer (1723–62) who first established a coordinate system, measured points, and derived a control network for the Moon. He then prepared two orthographic projection maps. It is surprising that no lunar coordinate system was established before this time, considering that maps of the Earth had contained latitude and longitude lines since the time of Ptolemy and many of the most beautiful maps ever produced came from cartographers in the sixteenth and seventeenth centuries. Most of the popular projections in use today were developed during this active period of cartography. I suspect that the early authors thought of their lunar drawings as illustrations, not as maps. Similarly, we view photographs as images rather than as cartography.

Tobias Mayer established a coordinate system for the Moon by solving for its axis of rotation and librations from measurements of the location of the crater Manilius relative to the lunar limb. He made these measurements with a glass micrometer that he designed and built. From these measurements, he determined the inclination of the lunar equator to the ecliptic to be 1°29' and the coordinates of Manilius to be latitude 14°34' and longitude 9°2'. The prime meridian was defined as the mean sub-Earth longitude. The positions of twenty-three points were measured relative to Manilius and their coordinates computed; then the coordinates of sixty-five additional points were estimated by referring their positions to the twenty-three points. I believe that this was the first control network of a planetary body. Maps prepared from these observations using orthographic projections were published after Mayer's death.

Johann H. Lambert (1728–77) made micrometer measurements of two hundred points and produced a lunar map in a modified stereographic projection; this map does not, however, have the positional accuracy or artistic beauty of Mayer's.

Johann H. Schröter (1745–1816) published *Selenotopographische Fragmente* in 1791 and 1802. In these two volumes he described in detail small regions of the Moon and drew elegant illustrations of them as they appeared under different illumination. For positional data he used Mayer's catalog of points.

William G. Lohrmann (1796–1840) measured the positions of seventy-nine points using a filar micrometer in 1822–6 and computed their coordinates. (The filar micrometer has threads or lines in the focal plane to aid in measuring small distances.) He then prepared a small map (38.5 cm in diameter), published in Leipzig in 1839. His large map (97.5 cm in diameter) was published long after his death by J. F. J. Schmidt in 1878.

Wilhelm Beer (1797–1850) and Johann H. Mädler (1794–1874) published a 97.5-cm-diameter map of the Moon in four sections between 1834 and 1836. For this project, Mädler measured the positions of 106 features using a filar micrometer. To determine the physical librations of the Moon, he used heliometer measurements of the small crater Mösting A. The heliometer is a double-image micrometer that was developed to measure the diameter of the Sun and was later used for other purposes, for example, in this case, measuring the distance on the Moon from a point to the limb.

Friedrick W. Bessel (1784–1846) had suggested Mösting A for the fundamental point; until this time the crater Manilius had been considered the reference point. Bessel (1839), and then his student, M. Wichmann (in 1844 and 1846), made heliometer measurements of Mösting A and derived a value of 1°32′9″ for the inclination of the lunar equator to the ecliptic. In 1878 Johann F. J. Schmidt (1825–84) published a map 194.9 cm in diameter divided into twenty-five sections. For control he relied on the points established by Lohrmann and Mädler. Illustrations and interesting descriptions of early lunar maps and their control can be found in Both (1960) and Kopal and Carder (1974).

With the introduction of photography, the measurement of points and the triangulation for coordinates of the points entered a new phase, making obsolete all previous work. Modern methods of measurement of points on photographs began with J. Franz in Germany and S. A. Saunder in England. Both made measurements relative to Mösting A, and Franz made heliometer measurements of eight secondary points to scale and orient the photographic plates. Saunder used twenty-six secondary points, of which six were in common with Franz. Franz measured five plates taken at the Lick Observatory and determined the coordinates of 150 points (Franz, 1899, 1901). Saunder measured four plates taken at the Paris Observatory and two taken at the Yerkes Observatory and published coordinates for 2885 points on the Moon (Saunder, 1911).

5.1.2. *Modern lunar control based on telescopic data*

The first modern lunar control network based on measurements of pictures taken via telescopes was published by G. Schrutka-Rechtenstamm (1958); it consisted of a recomputation of the measurements made by J. Franz for the coordinates of 150 points. This work is particularly important since, as Arthur (1968) pointed out, most of the telescopic networks depend on these data for scale and orientation of their pictures. Table 5.1 lists many of the most important networks, all of which go back to Franz's points for scale. They all use Mösting A as the fundamental point and, since the position of Mösting A had been determined relative to the limb, the coordinates had their origin at the center-of-figure. The center-of-figure, derived by fitting a circle to the limb, is an approximation to the center-of-mass, the correct origin for the coordinates. The path of an orbiting spacecraft is measured relative to the center-of-mass, so it was not until space exploration began that the proper origin was used.

Bessel (1839) selected the crater Mösting A as the fundamental point and proposed a method using a heliometer for determining its coordinates. Since that time, many series of heliometer measurements have been made. Koziel

Table 5.1. *Lunar networks derived from telescopic pictures.*

Number of points	Number of pictures	Reference
150	5	Franz, 1901
2,885	6	Saunder, 1911
150	5	Schrutka-Rechtenstamm, 1958
696	5	Baldwin, 1963
256	15	Marchant et al., 1964
196	8	Meyer and Ruffin, 1965
906	120	Mills and Sudbury, 1968
1,355	37	Arthur and Bates, 1968
1,156	10	Meyer, 1980

Table 5.2. *Coordinates of the fundamental lunar point, Mösting A.*

Latitude	Longitude	Radius (km)	Reference
−3°190	−5°172		Franz, 1899
−3.183	−5.172	1,741.770	Hayn, 1904
−3.180	−5.163	1,739.385	Schrutka-Rechtenstamm, 1956
−3.180	−5.164	1,738.733	Koziel, 1967
−3.222	−5.183	1,737.307	Schimerman, 1976 (Apollo)
−3.212	−5.211	1,737.527	Davies et al., 1987

(1967) reduced four series of 3,328 individual heliometer measurements made on 340 evenings over the period 1877 to 1915 to solve for the coordinates of Mösting A. Table 5.2 contains the coordinates of Mösting A as computed by various authors. Schimerman (1976) and Davies et al. (1987) refer coordinates to the center-of-mass origin. As observed by Arthur (1966), there is little variation in latitude and longitude; however, the radius was not well determined by early authors. Arthur correctly suggested that the radius was too large, at least if measured from the center-of-mass. Also included in Table 5.2 for comparison are the coordinates of Mösting A as determined in the Apollo Network (Schimerman, 1976) and the Unified Network (Davies et al., 1987). These networks are discussed in Section 5.4.1.

All of the lunar telescopic control networks have a center-of-figure origin for the coordinate system. The most recent and perhaps the best network is that of Meyer (1980). His project was designed to merge data with the higher-resolution Apollo mapping program and required the best possible photography. The measurements were made on ten glass plates taken over a period of twelve years with the 155 cm astrometric reflector of the U.S. Naval Observatory at Flagstaff, Arizona. This telescope was designed to take pictures with minimum geometric distortion, so accurate measurements can be made on the plates. Long exposures combined with the large aperture resulted in high-resolution images of excellent quality.

Figure 5.1
Mösting A as photographed by Lunar Orbiter.

The lunar telescopic control networks continue to be important because they contain the best positional data available over large areas of the Moon. The only better data are based on the Apollo mapping camera photography and Zond 6 and 8 images, which covered very limited regions.

5.1.3. *Mars control from telescopic data*

Telescopic observations of Mars began with Galileo in 1610, and before long observers were producing drawings of the red planet. Many of the familiar markings were recorded, and the coming and going of the polar caps recorded changing of the seasons on Mars. From drawings, maps of various projections were produced; Percival Lowell reproduced a number of these together with his own maps in *Mars and Its Canals* (1906). The most interesting drawings were those of Schiaparelli and Lowell depicting an extensive canal system. The debate regarding the possibility of canals on Mars continued for more than fifty years, although few measurements of points were made and the maps were little more than free-hand drawings. Lowell did paint a series of globes that are currently on display at the Lowell Observatory, Flagstaff, Arizona.

Gerard de Vaucouleurs of Harvard University and later the University of Texas carried out the Mars Map Project from 1958 to 1969 (de Vaucouleurs, 1969). The objective of the project was to use all the available ground-based data to map the surface of Mars; this material included both visual and photographic data and recorded changes to the surface markings with time. To carry out the mapping program it was necessary to derive a planetwide control network, which became a major part of the project.

The basic material for the project consisted of seventy-three drawings made by de Vaucouleurs in 1941 and thirty-two made in 1958. Also included were twenty-three composite photographs taken by Finsen in 1954 and fifty-three taken in 1956, and 233 taken by Leighton in 1956. Several hundred points were identified and measured on these pictures. In contrast to the Moon, where most points are topographic features (craters), on Mars, because of the great distance and low resolution, the points are albedo markings. A problem is that large albedo features change shape due to dust storms. Although this makes Mars an interesting planet, it also complicates the selection of unique, permanent measurable points.

Historical data, both visual and photographic, were incorporated in the Mars Map Project. This was particularly important because it was necessary to have a long time base in order to improve the siderial rotation period, which was used to determine longitudes of the points from transits. The datasets covered the time span from 1877 to 1969, almost one hundred years.

5.2. THE COORDINATE SYSTEMS

To define the coordinate system of a planet or satellite, it is first necessary to determine the direction of its axis of rotation; from this the equator is established and latitude is defined. An origin for the measurement of longitudes must be established, as well as conventions for positive and negative directions. Except for tradition, the definition of the prime meridian can be completely arbitrary.

A reference surface is required to develop the various cartographic projections used in mapping. Spheroids (ellipsoids of revolution) are defined for the Earth, Mars, and other planetary bodies, and simple spheres are used for Venus, Mercury, and other bodies that do not have significant polar flattening.

Conventions and definitions will be discussed first, then the method by which the rotational elements are related to the astronomical inertial coordinates system, and finally, the reference surfaces.

5.2.1. *Conventions and definitions*

Most of the lunar drawings of the seventeenth century, including the famous ones by Hevelius (1647) and Grimaldi and Riccioli (1651), showed north at the top. An exception was the Cassini (1692) map which had south at the top. We do not know why Cassini drew his map with south up; the most likely reason is that he used an inverting eyepiece, which was popular during that century.

Tobias Mayer established a coordinate system for the Moon in which north of the equator was positive latitude, south negative, and the 0° meridian was the mean sub-Earth meridian. East longitude was positive and west longitude negative, following the conventions on the Earth. This coordinate system is still in use today. However, after Mayer's maps were published with north at the top, all lunar maps were published with south at the top until the 1960s, when cartographers again placed north at the top.

The early maps of Mars were drawn much later than the lunar maps and were published with south at the top, as observed through telescopes with inverting eyepieces. Longitude went from 0° to 360° in the direction of increasing central longitude, as the planet is observed to rotate when viewed for an extended period with a telescope. The rotation of Mars is prograde, so west longitude was used. Thus, the longitude conventions for the Moon and Mars evolved from different traditions. With the advent of the space age, all maps of the Moon and Mars had north at the top and continued to do so as extensive mapping programs were started in both the United States and the Soviet Union.

The International Astronomical Union (IAU) met in Brighton, England, in August 1970. Mariners 6 and 7 had returned far-encounter and near-encounter pictures of Mars from their flybys. The first planetwide control network of Mars based on spacecraft data had just been completed (Davies and Berg, 1971). Mariners 8 and 9 were to be launched in 1971 to orbit Mars, and Mariner 10 was to be launched in 1973 to fly by Venus and Mercury. It was clear that the exploration of the Solar System with spacecraft was going to proceed at a rapid pace. To give guidance to the cartographers, Commission 16 of the IAU passed a resolution on planetographic coordinate systems containing two parts:

1. The rotational pole of a planet or satellite which lies on the north side of the invariable plane shall be called north, and northern latitudes shall be designated as positive.
2. The planetographic longitude of the central meridian, as observed from a direction fixed with respect to an inertial coordinate system, shall increase with time. The range of longitudes shall extend from 0° to 360°.

Thus, the traditional coordinate system of Mars was extended to all planets and satellites except the Earth and Moon.

At the IAU General Assembly in 1976, Commissions 4 and 16 established a Joint Working Group on Cartographic Coordinates and Rotational Elements of the Planets and Satellites. Since that time, this Working Group has addressed issues and conventions related to mapping the bodies of the Solar System. The rotational elements define the direction of the axis of rotation and the rate of rotation relative to an inertial coordinate system. Latitudes are defined by reference to the adopted axis of rotation, and longitudes by an arbitrarily chosen prime meridian whose position on the surface is specified, where possible, by a suitable observable feature. For example, the 0° longitude on Mars is defined by the small crater Airy-0 (de Vaucouleurs et al., 1973), and the 20° longitude on Mercury is defined by the small crater Hun Kal (Davies and Batson, 1975). For some of the planets and most of the satellites, it is sufficient to assume that the reference surface is spherical, but for others it is necessary to adopt an ellipsoid of revolution (defined by rotating an ellipse about its minor axis). The International Association of Geodesy (IAG) and the Committee on Space Research (COSPAR) expressed interest in the activities of the IAU Working Group, so in 1985 it became the IAU/IAG/COSPAR Working Group on Cartographic Coordinates and Rotational Elements of the Planets and Satellites.

5.2.2. *Rotational elements*

The rotational elements describe the coordinate systems with respect to an inertial coordinate system. The Working Group first reported to the 1979 IAU General Assembly (Davies et al., 1980) and described the coordinate systems with respect to the standard Earth equator and equinox of 1950.0, i.e., in effect an inertial coordinate system tied to the fundamental star catalog FK4. Variable quantities are expressed in units of ephemeris days (or Julian ephemeris centuries of 36,525 days) from the standard epoch of 1950 January 1.0, ephemeris time (ET), or Julian ephemeris date 2433282.5. The second report of the Working Group was presented to the 1982 IAU General Assembly (Davies et al., 1983); it gave the coordinate systems with respect to the standard equator and equinox of 1950.0 and also with respect to the new standard equator, equinox, and epoch of J2000, i.e., of 2000 January 1.5 or JD 2451545.0, which is now tied to the new star catalog FK5. The third report of the Working Group was presented to the 1985 IAU General Assembly (Davies et al., 1986); it contained data in the J2000 system only.

The direction of the north pole is specified by the value of its right ascension (α_o) and declination (δ_o), whereas the location of the prime meridian is specified by the angle W that is measured *along* the planet's equator in an easterly direction with respect to the planet's north pole *from* the node Q (located at right ascension $90° + \alpha_o$) of the planet's equator on the standard equator *to* the point B, where the prime meridian crosses the planet's equator (see Figure 5.2). The right ascension of the point Q is $90° + \alpha_o$, and the inclination of the planet's equator to the standard equator is $90° - \delta_o$. Because the prime

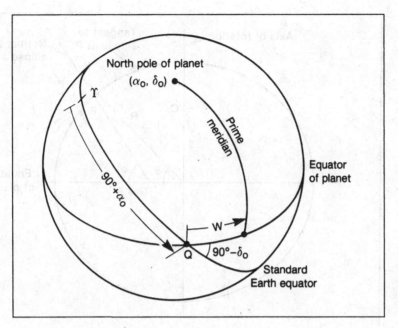

Figure 5.2
Reference system used to define
orientation of the planet.

meridian is assumed to rotate uniformly with the planet, W accordingly varies linearly with time. In addition, α_o, δ_o, and W may vary with time due to precession of the axis of rotation of the planet (or satellite). If W increases with time, the planet has a *direct* (or prograde) rotation, and if W decreases with time, the rotation is said to be *retrograde*.

To establish the coordinate system of a planet or satellite, it is necessary to determine the axis of rotation and rotational period. Sometimes these can be measured with telescopic or radar observations from the Earth or from spacecraft. Thus, the parameters for the Moon, Mars, and Jupiter were measured by telescope, and those of Mercury and Venus by radar. Light curves (created by measuring the total light flux with a photometer) have been used to estimate the rotation rates of the distant planets and some satellites. Dynamical studies are able to identify certain stable spin configurations (for instance, see Peale, 1977). Thus, most satellites (but not the Moon) in near orbits around major planets are in synchronous rotation with their spin axes normal to their orbital planes.

The angle W specifies the ephemeris position of the prime meridian, and for planets or satellites without any accurately observable fixed surface features, the adopted expression for W defines the prime meridian and is not subject to correction. Where possible, however, the cartographic position of the prime meridian is defined by a suitable observable feature, and so the constants in the expression $W = W_o + Wd$, where d is the interval in days from the standard epoch, are chosen so that the ephemeris position follows the motion of the cartographic position as closely as possible; in these cases, the expression for W may require emendation in the future. The most recent report of the

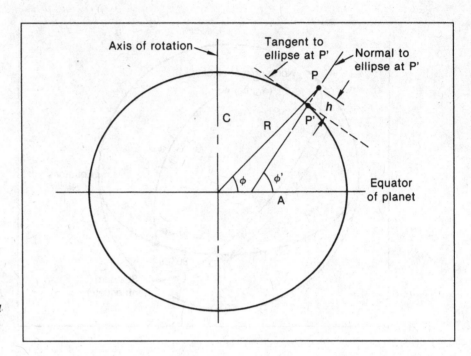

Figure 5.3
Planetocentric latitude (φ) and
planetographic latitude (φ') of
the point (P).

Working Group will contain the most up-to-date expressions for α_o, δ_o, and W.

5.2.3. Reference surfaces

In mapping, reference surfaces are used in transforming the coordinate system into a projection surface. The most common reference surfaces are spheres, ellipsoids of revolution, and three-axis ellipsoids. Recommended values for these reference surfaces are contained in the reports of the IAU Working Group and are modified frequently as new and improved measurements become available (see Davies et al., 1980, 1983, 1986).

If the reference surface is an ellipsoid of revolution (spheroid), it is necessary to distinguish between planetocentric latitude (φ) and planetographic latitude (φ') (see Figure 5.3).

If the equatorial radius is A and the polar radius C, then the polar flattening f is

$$f = \frac{A - C}{A} \tag{5.1}$$

Because h is small compared to R, it is usually sufficiently accurate to relate φ and φ' by

$$\tan \phi = \left(\frac{C}{A}\right)^2 \tan \phi' = (1 - f)^2 \tan \phi' \qquad (5.2)$$

The distinction between planetographic latitude ϕ' and planetocentric latitude ϕ is important because the latitude shown on maps is planetographic latitude, whereas the latitude used in most computations is planetocentric latitude.

5.3. ANALYTICAL PHOTOTRIANGULATION

Analytical phototriangulation is the computational procedure by which measurements of the locations of points on photographs are combined in blocks and surface coordinates (latitude and longitude) of the points determined. It is like a land survey, substituting photographs of the ground for walking on the ground. This method obviously has a great advantage when dealing with the planets and satellites. Most of the exploratory missions to the planets do contain imaging systems, and pictures taken by these cameras serve as the basic data source for the phototriangulation.

In this section the bundle method of phototriangulation will be discussed. The theory and methods were developed for using high-resolution aerial photographs of the Earth as a substitute for land surveys. There are many commercial companies and government agencies doing this work; the availability of powerful computers has contributed to the growth of the industry. Following the discussion of the theory is a section describing how the method is adapted to the planets and satellites.

5.3.1. *The bundle method*

An aerial photograph can be viewed as a bundle of light rays passing from the ground through the center-of-projection (the lens nodal point) to the film in the camera's focal plane (see Figure 5.4). There is a one-to-one correspondence between the points on the ground and points in the picture if the resolution and contrast on the film are adequate. Since the photograph is two-dimensional and the ground is three-dimensional, at least two pictures taken from different directions are required to define a model of the ground (see Figure 5.5). The picture might be taken on photographic film, by television (vidicon), by a charge-coupled device (CCD), or other imaging system.

Analytical photogrammetry has been under development for the past fifty years (see *Manual of Photogrammetry,* 1966) and is now based on a firm theoretical and practical footing. The modern treatment of analytical phototriangulation was developed by Duane Brown (1958), rigorously based on the theory of least squares, including weights and solutions for the standard

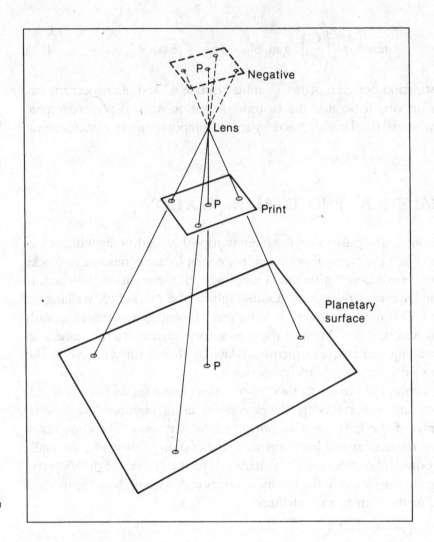

Figure 5.4
The aerial photograph can be thought of as a bundle of light rays passing from the ground through the center-of-projection (the lens) to the film.

errors of parameters. Since that time there have not been many theoretical changes, although there have been many improvements in computational algorithms, treatment of variables for particular problems, and the introduction of nonphotogrammetric measurements and constraints. Certainly the explosion in computer technology has added to the popularity and practicality of solving large systems of photogrammetric data. Ayeni (1982) has published a review and bibliography of phototriangulation that includes a classification of twenty-nine phototriangulation methods and describes thirty computer programs using analytical methods. The list is not comprehensive; many programs are not listed, and new programs are continually being developed for different applications. At RAND we have developed our own programs for vidicon imagery; on occasion we have used the SURBAT program of Duane Brown Associates and the GIANT program of the U.S. Geological Survey (Elassal, 1976).

A special class of aerial cameras is designed for obtaining pictures to be

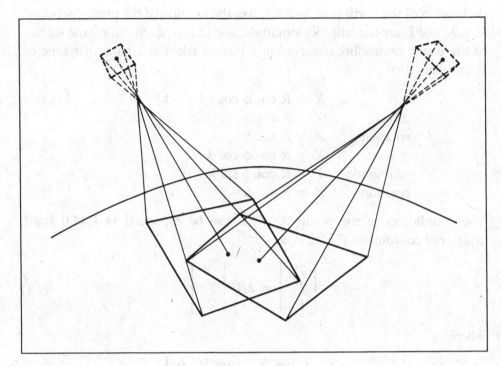

Figure 5.5
Two aerial photographs taken from different directions can define a three-dimensional surface.

used for photogrammetry and mapping. Typical mapping cameras have a 150-mm focal length lens, 23 × 23 cm format, and a between-the-lens shutter, so that the entire frame is exposed simultaneously. The lens is designed to minimize geometric distortion; 24-cm-wide film is fairly standard, and image motion compensation is usually included in the magazine design. Some cameras contain a glass plate reseau to make it possible to determine the extent of film expansion or shrinkage from measurements on the pictures.

A mapping camera was flown only on the Apollo 15, 16, and 17 lunar missions. The camera had a 75-mm focal-length lens and 115 × 115 mm format (Doyle, 1970). Also included in the payload was a stellar camera and an altimeter. Most U.S. planetary missions beyond the Moon with imaging experiments have flown with vidicon imaging systems. Many of these cameras have long focal lengths, so with the small format of the sensor (about 1 cm square), the photogrammetric problems and accuracy of solutions are somewhat different from those experienced with the conventional wide-angle mapping camera.

5.3.2. *Planetary analytical triangulation*

The coordinate systems of the planets and satellites and their relation to the J2000 inertial coordinate system was discussed in Section 5.2. The coordinate system of a planetary body is described by three equations (α_o, δ_o, W) that vary with time (see Figure 5.2); where α_o and δ_o are the right ascension and declination of the north pole, and W gives the location of the prime meridian. If ϕ, λ, and R are the latitude, longitude, and radius of a point on the surface of the planet or satellite observed in a picture taken at a particular time or Julian date, then

$$
\begin{aligned}
& X = R \cos \phi \cos (360° - \lambda) \\
\text{direct} \quad & \Upsilon = R \cos \phi \sin (360° - \lambda) \\
\text{rotation} \quad & Z = R \sin \phi \\
& X = R \cos \phi \cos \lambda \\
\text{retrograde} \quad & \Upsilon = R \cos \phi \sin \lambda \\
\text{rotation} \quad & Z = R \sin \phi
\end{aligned}
\tag{5.3}
$$

The coordinates of the points X, Υ, Z can be expressed in J2000 Earth equatorial coordinates P_x, P_y, P_z as

$$
\begin{bmatrix} P_x \\ P_y \\ P_z \end{bmatrix} = MV \begin{bmatrix} X \\ \Upsilon \\ Z \end{bmatrix}
\tag{5.4}
$$

where

$$
V = \begin{bmatrix} \cos W & -\sin W & 0 \\ \sin W & \cos W & 0 \\ 0 & 0 & 1 \end{bmatrix}
$$

and

$$
M = \begin{bmatrix} \cos (90° + \alpha_o) & -\sin (90° + \alpha_o) & 0 \\ \sin (90 + \alpha_o) & \cos (90° + \alpha_o) & 0 \\ 0 & 0 & 1 \end{bmatrix} \cdot
$$

$$
\begin{bmatrix} 1 & 0 & 0 \\ 0 & \cos (90° - \delta_o) & -\sin (90° - \delta_o) \\ 0 & \sin (90° - \delta_o) & \cos (90° - \delta_o) \end{bmatrix}
$$

then

$$
MV = \begin{bmatrix}
-\sin\alpha_o\cos W - \cos\alpha_o\sin\delta_o\sin W \\
\cos\alpha_o\cos W - \sin\alpha_o\sin\delta_o\sin W \\
\cos\delta_o\sin W
\end{bmatrix}
$$

$$
\begin{bmatrix}
\sin\alpha_o\sin W - \cos\alpha_o\sin\delta_o\cos W & \cos\alpha_o\cos\delta_o \\
-\cos\alpha_o\sin W - \sin\alpha_o\sin\delta_o\cos W & \sin\alpha_o\cos\delta_o \\
\cos\delta_o\cos W & \sin\delta_o
\end{bmatrix}
$$

The spacecraft location when the picture was taken, in J2000 Earth equatorial coordinates, is S_x, S_y, S_z; the origin is at the center of mass of the planet or satellite and is derived from the ephemeris and trajectory data. Translating the origin to the spacecraft, the coordinates ξ, η, ζ of the point in camera coordinates become

$$
\begin{bmatrix} \xi \\ \eta \\ \zeta \end{bmatrix} = C \begin{bmatrix} P_x - S_x \\ P_y - S_y \\ P_z - S_z \end{bmatrix} \tag{5.5}
$$

where

$$
C = \begin{bmatrix} -\sin\alpha\cos K - \cos\alpha\sin\delta\sin K & \cos\alpha\cos K - \sin\alpha\sin\delta\sin K & \cos\delta\sin K \\ \sin\alpha\sin K - \cos\alpha\sin\delta\cos K & -\cos\alpha\sin K - \sin\alpha\sin\delta\cos K & \cos\delta\cos K \\ \cos\alpha\cos\delta & \sin\alpha\cos\delta & \sin\delta \end{bmatrix}
$$

α, δ are the right ascension and declination of the optical axis, and K is the rotation angle in the picture plane measured from the trace of the J2000 Earth equatorial plane on the picture plane. The computed coordinates of the point on the camera detector relative to the intersection of the optical axis of a distortion-free lens with the detector are then

$$
x_c = \frac{\xi}{\zeta}f, \quad y_c = \frac{\eta}{\zeta}f \tag{5.6}
$$

where f is the focal length of the camera lens.

The location coordinates of the same point in the picture are measured by counting pixels and, using the reseau, the measurements are corrected for geometric distortions and scaled to millimeters on the faceplate of the television camera (x_o, y_o). The residuals are then $(x_c - x_o, y_c - y_o)$, the differences between the computed and measured coordinates; the analytical triangulation is designed to minimize the sum of the squares of those residuals by solving for corrections to selected parameters. Normally, the parameters to be improved are the latitude and longitude of the points (ϕ, λ) and the three angles of the C matrix (α, δ, K). The triangulation program permits solving for the radius assuming that all points lie on a sphere, or the three axes of a triaxial ellipsoid, or the radii of each point with no constraints. Another option includes solving for improvements in the direction of the spin axis (α_o, δ_o). Some of these combinations involve highly correlated parameters, so care must be exercised to obtain meaningful results. In the planetary analytical triangulation program, the S vector is not easily improved because of the small format of the television camera and the fairly long focal lengths. The planetary radii, R, and the S vector, S_x, S_y, S_z, are normally expressed in kilometers and the measurements x_o, y_o and the focal length, f, are normally in millimeters.

5.4. THE CONTROL NETWORKS

As the exploration of the Solar System by spacecraft is carried out, control networks of the solid surfaces of the planets and satellites are being developed as a part of the planetary mapping program. This section reviews the results from the computations of the control networks of the many bodies in the

Solar System. Also discussed are plans for further exploration by spacecraft and the expected contributions to the geodetic control networks of the bodies of the Solar System.

5.4.1. *Control networks of the Moon*

The lunar control networks are divided into regions of available photographic coverage. There is the front-side telescopic region, the Lunar Orbiter farside region, and the Apollo 15, 16, 17 and Zond 8 areas superimposed on the two large regions. The Apollo and Zond pictures are of better quality than the telescopic or Orbiter pictures and should give improved coordinates – the Apollo data certainly do.

Laser retroreflectors were placed on the surface of the Moon by Apollos 11, 14, and 15 and Lunokhod 2, and analysis of the lunar laser ranging data has led to very accurate determination of the coordinates of the retroreflectors (Bender et al., 1973; Ferrari et al., 1980). King et al. (1976) combined laser ranging data with very long baseline interferometry using the Apollo Lunar Surface Experiments Package (ALSEP) telemetry transmitters to determine very accurate coordinates of the Apollo 12, 14, 15, 16, and 17 ALSEP stations. The Apollo 15 retroreflector and the Apollo 15, 16, and 17 ALSEP transmitters were identified on high-resolution Apollo photographs and their positions transferred to mapping camera frames, thus permitting the comparison of these coordinates with the Apollo control network coordinates. For comparison with past work, the Defense Mapping Agency (DMA) Apollo Control Network coordinates of Mösting A are included in Table 5.2 (Schimerman, 1976). Also given are the coordinates from the Unified Control Network (Davies et al., 1987), in which the origin is defined by the lunar laser ranging experiment. The origin of the DMA solution is defined by the Apollo spacecraft orbital data.

The Moon is the only extraterrestrial body for which telescopic photographic plates have made a significant contribution to the establishment of a control network. Table 5.1 lists a few of the modern studies aimed at determining coordinates of features on the Moon; Franz (1901) and Saunder (1911) were pioneers in the use of photogrammetric methods to compute coordinates. The telescopic data are still important and in use (Meyer, 1980) because spacecraft have not yet yielded better materials over large regions of the near side.

Pictures taken by spacecraft have been the source materials for the control networks listed in Table 5.3. The Lunar Orbiters provided medium-resolution global coverage of the lunar surface. However, in order to make measurements on the pictures, it was necessary to reassemble the small framelets precisely to the film reseau – a time-consuming and laborious task. This camera system was not designed with photogrammetric applications in mind.

The Apollo 15, 16, and 17 orbiting spacecraft carried metric cameras specially designed and calibrated for mapping applications. Although the pho-

Table 5.3. *Control networks of the Moon based on pictures taken by spacecraft.*

Spacecraft	Number of points	Number of pictures	Reference
Lunar Orbiter	1,789	41	Catalog of Lunar Positions (1975)
Apollo 15	4,900	630	Schimerman et al. (1973)
Apollo 15, 16, 17	12,541		Schimerman (1976)
Apollo 15, 16, 17	5,324	1,244	Doyle et al. (1977)
Zond 6, 8	377		Bol'shakov et al. (1975)
Zond 6, 8	131	4	Ziman et al. (1975)
Zond 8	105	3	Chikmachev (1975)

tographic coverage was limited, the accuracy of the network in that region is much greater than in any other area of the Moon (Schimerman et al., 1973; Schimerman, 1976; Doyle et al., 1977). Zond 6 and 8 pictures were tied photogrammetrically to points on the nearside telescopic net, which extended the net into the eastern far side of the Moon (Bol'shakov et al., 1975; Ziman et al., 1975; Chikmachev, 1978). A unified control network of the Moon is being prepared – the near side has been completed (Davies et al., 1987), and the results will be extended to the far side. The origin (center-of-mass) of this system is defined by the laser ranging experiment; the Apollo network was transformed to this origin, and the telescopic network was transformed onto the Apollo network.

5.4.2. *Control networks of Mars and its satellites*

A planetwide control network of Mars was derived from pictures taken by the Mariner 6 and 7 spacecraft (Davies, 1972). Because the Mariners 6 and 7 were flyby missions, the datasets were small and the analytical triangulation included only 115 points and fifty-seven pictures. The Mariner 9 spacecraft took thousands of pictures during its active year in orbit and more than a thousand pictures were required to obtain full coverage of the surface of Mars. This mission was followed by the Viking missions, which returned more pictures by the thousands. Since the time of Mariners 6 and 7 there has been continuous updating of the control network.

The Geodesy/Cartography Group of the Mariner 9 Television Team selected the crater Airy–0 to define the 0° longitude on Mars (de Vaucouleurs et al., 1973). They also chose the same reference ellipsoid of revolution used in the Mariner 6 and 7 control network computations (3393.4 km equatorial, 3375.8 km polar). These values are still in use. The direction of the axis of rotation was analyzed and updated by the Viking Lander Radio Science Team (Mayo et al., 1977; Michael, 1979). The location of the Viking 1 lander was identified on high-resolution Viking Orbiter pictures (Morris and Jones, 1980), so the accurate coordinates of the lander site (Michael, 1979) could be incorporated

Table 5.4. *The control networks of Mars.*

Parameter	Davies and Arthur (1973)	Davies (1978)	Davies and Katayama (1983)	Unpublished (1989)
Points	1,205	3,037	6,853	9,292
Pictures	598	928	1,054	1,054
M9			757	1,425
Viking				
Measurements		17,224	47,524	62,072
Normal equations		8,858	19,139	26,021
Overdetermination		1.94	2.48	2.39
Standard error* (μm)		16.70	18.06	14.12

*Standard error of the residuals of the measurements at the focal plane of the vidicon cameras.

in the analytical triangulation. This tie to the surface is very important. Without it, the entire triangulation is dependent on the orbital positions of the spacecraft and statistical solutions for the camera pointing. In the region around the Viking 1 lander the positional accuracy of the coordinates of control points is better than 100 m; in contrast, the accuracy of most coordinates on Mars is 1 to 6 km. Unfortunately, the location of the Viking 2 lander could not be identified on Viking orbiter pictures, so its coordinates could not be incorporated in the analytical triangulation.

The first published Mariner 9 control network (Davies and Arthur, 1973) was broken into five overlapping regions, each region computed independently. Each region contained almost the maximum number of parameters permitted by the analytical triangulation program and the available computer. The regions contained 273 points, 141 pictures; 335 points, 140 pictures; 363 points, 140 pictures; 231 points, 142 pictures; and 238 points, 154 pictures. In contrast, the most recently published horizontal control network of Mars contained 6,853 points and 1,811 pictures (Davies and Katayama, 1983a). The growth in the Mars network is shown in Table 5.4.

Vertical control on Mars has been derived from many sources (Wu, 1978); these include Mariner 9 and Viking Orbiter 1 and 2 radio occultations, ground-based radar measurements using the Haystack and Goldstone radio telescopes, and measurements by the Mariner 9 ultraviolet spectrometer, infrared interferometer spectrometer, and infrared radiometer and photogrammetry (see Chapter 4).

The pictures of Mars taken by the Soviet spacecraft Mars 4 and 5 were used to establish a control network of the region covered by the photography (Tjuflin et al., 1980). The phototriangulation was tied to the Mariner 9 control network (Davies and Arthur, 1973), so the origin of longitudes was the crater Airy–0. The network contained 184 new points and thirteen Mariner 9 points measured on the Mars 4 and 5 pictures.

The first pictures of the satellites of Mars, Phobos, and Deimos, were taken by the Mariner 9 spacecraft. A control network of Phobos was computed based on measurements of thirty-eight points on nine pictures (Duxbury, 1974). A best-fit triaxial ellipsoid was computed by Duxbury (1974), using the radii at the thirty-eight points; the three axes were $a = 13.5 \pm 0.5$ km, $b = 10.8 \pm 0.7$ km, and $c = 9.4 \pm 0.7$ km, where a is in the direction of Mars, b in the orbit plane, and c normal to the orbit plane. Duxbury (1977) determined the axes of triaxial ellipsoids of Phobos and Deimos by overlaying various shapes onto the Mariner 9 pictures. Using this technique, he obtained values of 13.5 km, 11.5 km, and 9.5 km for the three Phobos radii and 7.5 km, 6.1 km, and 5.5 km for the three Deimos radii.

5.4.3. Control network of Mercury

The Mariner 10 spacecraft was launched November 3, 1973, flew by Venus on February 5, 1974, and passed by the dark side of Mercury on March 29, 1974. The dark-side pass was chosen to obtain an occultation so that a sensitive search for an atmosphere could be made and to obtain an accurate measurement of the planetary radius. Mosaics of the surface were obtained by the television camera during approach and when departing the planet.

As the Mariner 10 spacecraft went around the Sun in its orbit once, Mercury went around the Sun on its orbit twice, and they met again on September 21, 1974. This time a bright-side flyby was made and mosaics of most of the illuminated surface were obtained. A third encounter was made March 16, 1975; again, a dark-side pass was chosen to confirm the intrinsic magnetic field of Mercury.

The horizontal control network of Mercury was computed photogrammetrically from pictures taken by the Mariner 10 spacecraft on its three encounters with Mercury (Davies and Katayama, 1976). All of the control points were constrained to lie on a sphere with a radius of 2,439 km. This radius is consistent with the two occultation measurements made when the Mariner 10 spacecraft passed behind Mercury (Howard et al., 1974) and the measurements made from ground-based radar (Ash et al., 1971). The spin axis was assumed to be normal to the orbital plane. The control network contained 2,378 points and 25,504 measurements on 788 pictures; there were 435 frames from Mercury 1, 333 from Mercury 2, and twenty from Mercury 3. The small crater Hun Kal defined the 20° meridian on Mercury, establishing the system of longitudes. The parameters for the phototriangulation are summarized in Table 5.5.

5.4.4. Control networks of the satellites of Jupiter

Voyager 2 was launched on August 20, 1977, and Voyager 1 followed on September 5, 1977. These spacecraft were designed to fly by Jupiter and use the gravity assist of that planet to bend their trajectories so they would en-

Table 5.5. *The control networks of Mercury.*

Parameter	Davies and Batson (1975)	Davies (1976)	Davies and Katayama (1976)	Unpublished (1989)
Points	1,328	1,774	2,378	2,393
Pictures	545	680	788	811
Measurements	11,234	16,148	25,504	26,142
Normal equations	4,291	5,588	7,120	7,219
Overdetermination	2.62	2.89	3.58	3.62
Standard error[a] (μm)	58.92	28.36	23.11	21.93

[a] Standard error of the residuals of the measurements at the focal plane of the vidicon cameras.

counter Saturn. Thus, their aiming points at Jupiter were defined by their time of arrival.

Voyager 1 passed by Jupiter on March 5, 1979, and took many pictures of the planet and its four large satellites, Io, Europa, Ganymede, and Callisto. Voyager 1 was targeted to make a close pass by Io because telescopic studies suggested it possessed some unusual properties. Voyager 2 flew by Jupiter on July 9, 1979, and was targeted to fly outside the orbit of Europa and thus to avoid much of the radiation trapped in Jupiter's magnetic field.

The Voyager 1 and 2 spacecraft pictures taken of the four large Galilean satellites of Jupiter as they flew by the planet were used in phototriangulations to measure the mean radii of these satellites as well as to determine the coordinates of the control points (Davies and Katayama, 1981). Because these were flyby missions, the resolution of the pictures of these bodies varied greatly with longitude. The direction of the spin axis of the satellites has been assumed to be normal to their orbital planes.

The 1981 phototriangulation of Io contained 504 control points that were measured on 191 Voyager 1 and forty-three Voyager 2 pictures. The mean radius of Io was measured as 1,815 km. Longitudes on Io are defined by the *W* equation.

Active volcanism was discovered on Io and estimated resurfacing rates were sufficient to account for the lack of observed impact craters on Io's surface (Johnson et al., 1979). If a small surface feature were selected to define the longitude system on Io, there is no assurance that it would be recognized when the Galileo spacecraft views Io in 1995. Thus, the selection of a surface feature for the definition of longitudes has been postponed.

The 1981 phototriangulation of Europa contained 112 control points that were measured on fifty-three Voyager 1 and sixty-two Voyager 2 pictures. The mean radius of Europa was measured as 1,569 km. The 182° meridian passes through the center of the crater Cilix, thus defining the system of longitudes.

The 1981 phototriangulation of Ganymede contained 1,547 control points that were measured on 145 Voyager 1 and 137 Voyager 2 pictures. The mean

Table 5.6. *Control networks of the satellites of Jupiter.*

Parameter	Io Davies and Katayama (1981)	Io Unpublished (1989)	Europa Davies and Katayama (1981)	Europa Unpublished (1989)	Ganymede Davies and Katayama (1981)	Ganymede Unpublished (1989)	Callisto Davies and Katayama (1981)	Callisto Unpublished (1989)
Points	504	712	112	179	1,547	1,884	439	818
Pictures	234	252	115	120	282	302	200	286
Measurements	8,510	12,356	2,564	3,848	14,724	25,780	6,670	15,786
Normal equations	1,710	2,180	569	718	3,940	4,674	1,478	2,494
Overdetermination	4.98	5.67	4.45	5.36	3.74	5.52	4.51	6.33
Standard error[a] (μm)	18.27	11.02	14.88	9.93	22.58	10.50	19.75	11.40

[a] Standard error of the residuals of the measurements at the focal plane of the vidicon cameras.

radius of Ganymede was measured as 2,631 km. The 128° meridian passes through the center of the crater Anat.

The 1981 phototriangulation of Callisto contained 439 control points that were measured on 119 Voyager 1 and eighty-one Voyager 2 pictures. The mean radius of Callisto was measured as 2,400 km. The 326° meridian passes through the center of the crater Saga.

The first control networks of the Galilean satellites were published in 1981 (Davies and Katayama, 1981). Because of the upcoming Galileo mission to Jupiter, there has been continued interest in improved positional data for the satellites. The more recent work has not been published; however, results are reviewed in Table 5.6 and can be compared with the 1981 computations.

5.4.5. *Control networks of the satellites of Saturn*

After their Jupiter encounters, the Voyager spacecraft continued on the long flight to Saturn. The strategy was to target Voyager 1 for a close flyby of Saturn's large satellite, Titan. If the data from Titan were returned successfully, then the aiming point for Voyager 2 would send it on to Uranus. If the Titan encounter was not successful, then Voyager 2 would be retargeted to Titan.

Voyager 1 flew by Saturn November 12, 1980, and returned beautiful data from Titan. Titan was completely cloud covered; we could not see the surface; we could not map Titan. Voyager 2 encountered Saturn on August 25, 1981, and continued on a trajectory to Uranus.

The Voyager 1 and 2 spacecraft took pictures of the small satellites of Saturn as they flew through the complex system. Many fewer pictures of these satellites were taken than those of Jupiter, because the communication data rate was lower and because there were so many diverse targets. Voyager 1 did take useful pictures of Mimas, Dione, and Rhea for phototriangulation; and Voyager 2 took pictures of Enceladus, Tethys, and Iapetus. The mean radii of these satellites were computed in the phototriangulations. To establish their coordinate systems, it was assumed that their spin axes were normal to their orbital planes.

The phototriangulation of Mimas contained 110 control points that were measured on thirty-two Voyager 1 frames (Davies and Katayama, 1983b). The mean radius of Mimas was measured as 197 km. If Mimas were in hydrostatic equilibrium, it would be ellipsoidal with $a - c$ of 16 to 20 km (Dermott, 1979, 1984; Zharkov et al., 1985). When these axes were treated as free parameters in the analytical triangulation, $a - c$ was measured as 6 km (Davies and Katayama, 1983b). Models of the interior composition of Mimas indicate that the value of $a - c$ should lie between 11 and 20 km (Dermott, 1984; Zharkov et al., 1985), suggesting that the value of 6 is unrealistically small. Limb measurements by Dermott and Thomas (1988) yield an $a - c$ of 17 km, no doubt a more accurate figure. The 162° meridian passes through the center of the crater Palomides, thus defining the system of longitudes.

The phototriangulation of Enceladus contained seventy-one control points that were measured on twenty-two Voyager 2 pictures (Davies and Katayama, 1983b). The mean radius of Enceladus was measured as 251 km. If Enceladus were in hydrostatic equilibrium, it would be ellipsoidal with $a - c$ about 11 km (Dermott, 1979, 1984). When these axes were treated as free parameters in the analytical triangulation, $a - c$ was measured as 8 km (Davies and Katayama, 1983b). This measurement would be consistent with a satellite composed primarily of ice. The 5° meridian passes through the center of the crater Salih.

The phototriangulation of Tethys contained 110 control points that were measured on six Voyager 1 and twenty-one Voyager 2 frames (Davies and Katayama, 1983c). The mean radius of Tethys was measured as 524 km. The 299° meridian passes through the center of the crater Arete.

The phototriangulation of Dione contained 126 control points that were measured on twenty-seven Voyager 1 pictures and one Voyager 2 picture (Davies and Katayama, 1983c). The mean radius of Dione was measured as 559 km. The 63° meridian passes through the center of the crater Palinurus.

The phototriangulation of Rhea contained 352 control points that were measured on eighty-one Voyager 1 and three Voyager 2 pictures (Davies and Katayama, 1983d). The mean radius of Rhea was measured as 764 km. The 340° meridian passes through the center of the crater Tore.

The phototriangulation of Iapetus contained sixty-two points that were measured on fourteen Voyager 1 and sixty-six Voyager 2 frames (Davies and Katayama, 1984). The mean radius of Iapetus was measured as 718 km. The 276° meridian passes through the center of the crater Almeric, thus defining the system of longitudes.

Phototriangulation parameters for the satellites of Saturn are given in Table 5.7.

Table 5.7. *Control networks of the satellites of Saturn.*

Parameter	Mimas[a]	Enceladus[a]	Tethys[b]	Dione[b]	Rhea[c]	Iapetus[d]
				l		
Points	110	71	110	126	352	62
Pictures	32	22	27	28	84	80
Measurements	1,356	994	924	1,322	3,814	1,872
Normal equations	316	208	302	337	957	365
Overdetermination	4.26	4.78	3.06	3.92	3.99	5.13
Standard error[e] (μm)	13.43	19.19	12.15	13.94	11.16	11.71

[a]Davies and Katayama (1983b).
[b]Davies and Katayama (1983c).
[c]Davies and Katayama (1983d).
[d]Davies and Katayama (1984).
[e]Standard error of the residuals of the measurements at the focal plane of the vidicon cameras.

5.4.6. *Control networks of the satellites of Uranus*

The Voyager 2 spacecraft took pictures of the satellites of Uranus as it explored this strange planetary system. The equator of the planet and the orbital planes of the satellites are nearly perpendicular to Uranus' orbital plane, forming a configuration unique in the solar system. At the time of the encounter, the southern hemispheres of Uranus and the satellites were illuminated by the Sun, thus limiting the areas that could be mapped. Equatorial lighting will be available in the year 2007. Pictures of the south polar regions of five satellites, Miranda, Ariel, Umbriel, Titania, and Oberon, were taken during the flyby and were used to compute control networks (Davies et al., 1987). Coordinate systems were established by assuming that their axes of rotation were normal to their orbital planes.

Because the approach trajectory and the Sun were approximately from the south polar direction, it was not possible to extend the coverage with approach pictures. Thus, all of the pictures of the satellites covered only their southern hemispheres. The spacecraft did pass near Miranda, and a beautiful high-resolution mosaic was obtained.

The phototriangulation of Miranda was based on 796 measurements of 103 control points on thirteen pictures. The shape of Miranda is a triaxial ellipsoid with radii 241 km, 235 km, and 232 km. The phototriangulation of Ariel incorporated measurements of fifty-two points on ten pictures. The radius of Ariel is 579 km. The phototriangulation of Umbriel contained forty-three points and only six pictures; the radius was computed to be 586 km. The phototriangulation of Titania contained measurements of forty-six control points on twenty pictures. The radius of Titania is 790 km. The phototriangulation of Oberon was based on measurements of thirty-four points on five pictures. The radius was computed to be 762 km (Davies et al., 1987).

Table 5.8. *Control networks of the satellites of Uranus.*

Parameter	Miranda	Ariel	Umbriel	Titania	Oberon
Points	103	52	43	46	34
Pictures	13	10	6	20	5
Measurements	796	480	216	1,070	316
Normal equations	249	134	104	152	83
Overdetermination	3.20	3.58	2.07	7.04	3.81
Standard error[a] (μm)	13.00	12.55	14.20	12.93	10.63

[a]Standard error of the residuals of the measurements at the focal plane of the vidicon cameras.
From Davies et al. (1987).

A summary of the phototriangulation parameters of the satellites of Uranus is given in Table 5.8.

5.4.7. *Control networks of Triton, the large satellite of Neptune*

The Voyager 2 spacecraft flew by Neptune in August 1989 and took many pictures of the planet and its large satellite Triton. The pictures included high-resolution mosaics. Although it was found that Triton has a thin atmosphere, surface features could be seen in all the pictures. Computations of the control network started immediately.

The control network of Triton contains eighty points measured on fifty-eight pictures. There are 335 normal equations with 904 measurements; the overdetermination is 2.70. The standard error of the residuals is 14.01 μm, and the radius of Triton was computed to be 1,354 km.

5.5. FUTURE MISSIONS AND PLANS

The Magellan spacecraft was launched March 5, 1989, on a trajectory that will take it to Venus. It will arrive on August 10, 1990, go into orbit, and spend 243 days mapping most of the surface of Venus with pictures taken by a synthetic aperature radar. Analysis of the radar data will lead to improved coordinates of features on the surface of Venus. A preliminary control network has been developed based on the Soviet Venera 15 and 16 maps and high-resolution radar pictures acquired by the Arecibo and Goldstone antennas.

The Galileo spacecraft was launched on October 18, 1989 in the direction of Venus. After flying by Venus in February 1990, it will return to Earth in December 1990 and leave on a trajectory that will carry it past Mars' orbit into the asteroid belt. Galileo will return to Earth in December 1992 and depart on a trajectory to Jupiter; it will arrive in December 1995. The spacecraft will release a probe into the atmosphere, then burn its large engine and go into orbit. During the next two years, Galileo will make ten orbits around Jupiter, with many close encounters of Europa, Ganymede, and Callisto.

Mars will be the target of extensive exploration during the next two decades.

The U.S. Mars Observer spacecraft will be launched in 1992, followed by the Soviet 1994 spacecraft. Before the end of that decade, rovers will be exploring the surface of Mars, and perhaps rock samples will be returned to Earth for analysis.

Asteroids and comets will become interesting targets for exploration, and both Soviet and U.S. spacecraft will be visiting some of them in the next decade. The comet Halley was explored by Soviet and European spacecraft during its 1986 rendezvous with the inner Solar System. The Galileo spacecraft is expected to make fast flybys of one or two asteroids on its extended flight to Jupiter.

5.6.   SUMMARY

The geodetic control network of a planetary body is similar to a land survey on the Earth; it consists of a list of accurate coordinates of defined points. On the planets the points are identified and measured on pictures usually taken by spacecraft. The latitude and longitudes of the points are computed by photogrammetric methods similar to those developed to extend control on Earth with aerial photographs. The control network is important to the positioning of the latitude/longitude grid on maps of the planet or satellite.

The first control network of the Moon was computed by Tobias Mayer from measurements of points made at the telescope. Fredrick Bessel suggested that Mösting A be the fundamental point and all other points be measured relative to it. F. Franz in Germany and S. A. Saunder in England were the first to measure photographs taken with telescopes. The Moon is the only planetary body for which telescopic pictures are still useful for control. Pictures taken by the Lunar Orbiter, Apollos 15, 16, and 17, and Zonds 6 and 8 have contributed to the lunar control network. Although efforts are being made to combine and unify control on the Moon, the database is inadequate to meet future requirements.

Mariner 9 and the Viking missions have contributed to the development of a control network that covers the entire surface of Mars. The horizontal accuracies of the coordinates of points in the network vary greatly. In the area around the Viking lander, the accuracy is very good, about 100 m. Leaving this area are high-resolution strips of Viking pictures that encircle the planet. The accuracy of coordinates along the strips varies greatly, from about 500 m to 4 km. Coordinates of points outside the strips are known to 5 or 6 km. The new Mars missions should contribute to improving the accuracy of the coordinates.

The Mariner 10 mission to Mercury and the Voyager missions to the outer planets were the first explorations of these planets and satellites, using framing cameras. These "first looks" contributed greatly to our understanding of the bodies and permitted us to produce initial control networks. Because all the encounters were flyby (in contrast to orbital), there are serious gaps in the coverage and, hence, the control networks. Future missions like Galileo should contribute a great deal to improving the accuracy of the coordinates and, of course, the coverage of the network.

The International Astronomical Union, the International Association of Geodesy, and the Committee on Space Research established a Working Group on Cartographic Coordinates and Rotational Elements of the Planets and Satellites to define coordinate systems and mapping conventions for the planets and satellites. This Working Group sets standards that are used in the production of maps by all the countries involved in the exploration of the Solar System.

5.7. REFERENCES

American Society of Photogrammetry. 1966. *Manual of Photogrammetry,* 3rd ed. [See Whitmore, G. D., and Thompson, M. M., pp. 1–16, and others.]

Arthur, D. W. G. 1966. The validity of selenodetic positions. *Communications of the Lunar and Planetary Laboratory,* vol. 5, part 3. The University of Arizona, pp. 19–30.

Arthur, D. W. G. 1968. A new secondary selendotic triangulation. *Communications of the Lunar and Planetary Laboratory,* vol. 7, part 5. The University of Arizona, pp. 303–12.

Arthur, D. W. G., and Bates, P. 1968. The Tucson selenodetic triangulation. *Communications of the Lunar and Planetary Laboratory,* vol. 7, part 5. The University of Arizona, pp. 313–60.

Ash, M. E., Shapiro, I. I., and Smith, W. B. 1971. The system of planetary masses. *Science* 174 (4009): 551–6.

Ayeni, O. O. 1982. Phototriangulation: a review and bibliography. *Photogrammetric Engineering and Remote Sensing* 48 (11): 1733–59.

Baldwin, R. B. 1963. *The Measure of the Moon.* Chicago: University of Chicago Press, pp. 212–45.

Bender, P. L., Currie, D. G., et al. 1973. The lunar laser ranging experiment. *Science* 182 (4109): 229–38.

Bessel, F. W. 1839. Ueber die Bestimmung der Libration des Mondes. *Astron. Nachr.* 16: 257.

Bol'shakov, V. D., Krasnopevtsev, B. C., et al. 1975. Photographic experiments on the AMS "Zond 6, 7, and 8." In Iu. I. Efrevnov, ed., *Atlas of the Far Side of the Moon,* part 3. Moscow: Nauka, pp. 20–51 (in Russian).

Both, E. E. 1960. *A History of Lunar Studies.* Buffalo: Buffalo Museum of Science.

Brown, D. C. 1958. *A Solution to the General Problem of Multiple Station Analytical Stereotriangulation.* RCA-MTP Data Reduction Technical Report No. 43, Patrick Air Force Base, Florida.

Catalog of Lunar Positions. 1975. Based on the Lunar Positional Reference System (1974). St. Louis AFS, Mo.: Defense Mapping Agency, Aerospace Center.

Chikmachev, V. I. 1978. A list of spatial coordinates of points on the lunar far side in the Mars oriental's region. *Astronomical Newsletter,* No. 986: pp. 1–3 (Published by the Bureau of Astronomical Information of the USSR Academy of Sciences) (in Russian).

Davies, M. E. 1972. Coordinates of features on the Mariner 6 and 7 pictures of Mars. *Icarus* 17 (1): 116–67.

Davies, M. E. 1976. *The Control Net of Mercury: January 1976.* The RAND Corporation, R–1914-NASA.

Davies, M. E. 1978. The control net of Mars: May 1977. *J. Geophys. Res.* 83 (B5): 2311–12.

Davies, M. E., Abalakin, V. K., et al. 1980. Report of the IAU Working Group on Cartographic Coordinates and Rotational Elements of the Planets and Satellites. *Celestial Mechanics* 22: 205–30.

Davies, M. E., Abalakin, V. K., et al. 1983.

Report of the IAU Working Group on Cartographic Coordinates and Rotational Elements of the Planets and Satellites: 1982. *Celestial Mechanics* 29: 309–21.

Davies, M. E., Abalakin, V. K., et al. 1986. Report of the IAU/IAG/COSPAR Working Group on Cartographic Coordinates and Rotational Elements of the Planets and Satellites: 1985. *Celestial Mechanics* 39: 103–13.

Davies, M. E., and Arthur, D. W. G. 1973. Martian surface coordinates. *J. Geophys. Res.* 78 (20): 4355–94.

Davies, M. E., and Batson, R. M. 1975. Surface coordinates and cartography of Mercury. *J. Geophys. Res.* 80 (17): 2417–30.

Davies, M. E., and Berg, R. A., 1971. A Preliminary Control Net of Mars. *J. Geophys. Res.* 76 (2): 373–93.

Davies, M. E., Colvin, T. R., Katayama, F. Y., and Thomas, P. C. 1987. The control networks of the satellites of Uranus. *Icarus* 71: 137–47.

Davies, M. E., Colvin, T. R., and Meyer, D. L. 1987. A unified lunar control network: the near side. *J. Geophys. Res.* 92 (B13): 14177–84.

Davies, M. E., and Katayama, F. Y. 1976. *The Control Net of Mercury: November 1976*. The RAND Corporation, R–2089-NASA.

Davies, M. E., and Katayama, F. Y. 1981. Coordinates of features on the Galilean satellites. *J. Geophys. Res.* 86 (A10): 8635–57.

Davies, M. E., and Katayama, F. Y. 1983a. The 1982 control network of Mars. *J. Geophys. Res.* 88 (B9): 7503–4.

Davies, M. E., and Katayama, F. Y. 1983b. The control networks of Mimas and Enceladus. *Icarus* 53 (2): 332–40.

Davies, M. E., and Katayama, F. Y. 1983c. The control networks of Tethys and Dione. *J. Geophys. Res.* 88 (A11): 8729–35.

Davies, M. E., and Katayama, F. Y. 1983d. The control network of Rhea. *Icarus* 56 (3): 603–10.

Davies, M. E., and Katayama, F. Y. 1984. The control network of Iapetus. *Icarus* 59 (2): 199–204.

de Vaucouleurs, G. 1969. *Research Directed Toward Development of a Homogeneous Martian Coordinate System*. Bedford, Mass.: Air Force Cambridge Research Laboratories, AFCRL–69–0507.

de Vaucouleurs, G., Davies, M. E., and Sturns, F. M., Jr. 1973. The Mariner 9 areographic coordinate system. *J. Geophys. Res.* 78 (20): 4395–404.

Dermott, S. F. 1979. Shapes and gravitational moments of satellites and asteroids. *Icarus* 37 (3): 575–86.

Dermott, S. F. 1984. Rotation and the internal structures of the major planets and their inner satellites. *Phil. Trans. R. Soc. Land* A313: 123–39.

Dermott, S. F., and Thomas, P. C. 1988. The shape and internal structure of Mimas. *Icarus* 73: 25–65.

Doyle, F. J. 1970. Photographic systems for Apollo. *Photogrammetric Engineering* 36 (10): 1039–44.

Doyle, F. J., Elassal, A. A., and Lucas, J. R. 1977. *Selenocentric Geodetic Reference System*. NOAA Technical Report NOS 70 NGS 5. Rockville, Md.: U.S. Department of Commerce.

Duxbury, T. C. 1974. Phobos: control network analysis. *Icarus* 23 (2): 290–9.

Duxbury, T. C. 1977. Phobos and Deimos: geodesy. In J. A. Burns, ed., *Planetary Satellites*. University of Arizona Press, pp. 346–62.

Elassal, A. A. 1976. *General Integrated Analytical Triangulation (GIANT)*. Reston, Va.: USGS Topographic Division.

Ferrari, A. J., et al. 1980. Geophysical parameters of the Earth–Moon system. *J. Geophys. Res.* 85 (B7): 3939–51.

Franz, J. 1899. Die Figur deo Mondes. *Astr. Beob.* (Konigsberg) 38: 1–34.

Franz, J. 1901. Ortsbestimmung von 150 Mondkratern. *Mitt. Sternwarte Breslau, I.*

Hayn, F. 1904. *Selenographische Koordinaten*. Leipzig: Sachs. Gesell. der Wiss., II Abh.

Howard, H. T., Tyler, G. L., et al. 1974. Mercury: results on mass, radius, ionosphere, and atmosphere from Mariner 10 dual-frequency radio signals. *Science* 185 (4146): 179–80.

Johnson, T. V., Cook, A. F., II, et al. 1979. Volcanic resurfacing rates and implications for volatiles on Io. *Nature* 280 (5725): 746–50.

King, R. W., Counselman, C. C., III, and

Shapiro, I. I. 1976. Lunar dynamics and selenodesy: results from analysis of VLBI and laser data. *J. Geophys. Res.* 81 (35): 6251–6.

Kopal, Z. 1966. *An Introduction to the Study of the Moon*. Dordrecht: D. Reidel.

Kopal, Z., and Carder, R. W. 1974. *Mapping of the Moon*. Dordrecht/Boston: D. Reidel.

Koziel, K. 1967. Recent researches on the determination of the Moon's physical libration constants. In Z. Kopal and C. L. Goudas, eds., *Measure of the Moon*. Dordrecht: D. Reidel.

Lowell, P. 1906. *Mars and its Canals*. London: Macmillan.

Marchant, M. Q., Hardy, M., and Breece, S. 1964. *Horizontal and Vertical Control for Lunar Mapping, Part Two: AMS Selenodetic Control System 1964*. Army Map Service Technical Report No. 29. Washington, D.C.: U.S. Army Map Service.

Mayer, T. 1971. Tobias Mayer's *Opera Inedita*. Eric G. Forbes, transl. New York: American Elsevier.

Mayo, A. P., Blackshear, W. T., et al. 1977. Lander locations, Mars physical ephemeris, and Solar System parameters: determination from Viking Lander tracking data. *J. Geophys. Res.* 82 (28): 4297–303.

Meyer, D. L. 1980. *Selenocentric Control System (1979)*. DMA TR 80–001, Defense Mapping Agency.

Meyer, D. L., and Ruffin, B. W., 1965. *Coordinates of Lunar Features, Group I and II Solutions*. ACIC Technical Paper No. 15.

Michael, W. H., Jr. 1979. Viking Lander tracking contributions to Mars mapping. *The Moon and the Planets* 20: 149–52.

Mills, G. A., and Sudbury, P. V. 1968. Absolute coordinates of lunar features III. *Icarus* 9 (3): 538–61.

Morris, E. C., and Jones, K. L. 1980. Viking 1 Lander on the surface of Mars: revised location. *Icarus* 44 (1): 217–22.

Peale, S. J. 1977. Rotation histories of the natural satellites. In J. A. Burns, ed., *Planetary Satellites*. Tucson: University of Arizona Press, pp. 87–112.

Saunder, S. A. 1911. The determination of selenographic positions and the measurement of lunar photographs. *Memoirs Royal Astron. Soc.* 60: 1–81.

Schimerman, L. A. 1976. *The Expanding Apollo Control System*. St. Louis, Mo.: Defense Mapping Agency, Aerospace Center.

Schimerman, L. A., Cannell, W. D., and Meyer, D. L. 1973. *Relationship of Spacecraft and Earthbased Selenodetic Systems*. St. Louis, Mo.: Defense Mapping Agency, Aerospace Center.

Schrutka-Rechtenstamm, G. 1956. Newreduktion der acht von J. Franz und der vier von F. Hayn gemessenen Mondkraterpositionen. *Sitz. Osten. Akad. Wiss. Math.-Naturw. Kl.*, Abt. II, 165 bd.

Schrutka-Rechtenstamm, G. 1958. Neureduktion der 150 Mondpunkte der Breslauer Messungen von J. Franz, *Sitz. Osterr. Akad. Wiss. Math.-Naturw. Kl.*, Abt. II, 167 Bd.

Tjuflin, Y. S., et al., 1980. Photogrammetric and cartographic processing of materials photographed from AMA Mars–4 and Mars–5. In Y. I. Efremov, ed., *The Surface of Mars*. Moscow: Nauka, pp. 82–92 (in Russian).

Wu, S. S. C. 1978. Mars synthetic topographic mapping. *Icarus* 33(3): 417–40.

Zharkov, V. N., Leontjev, V. V., and Kozenko, A. V. 1985. Models, figures, and gravitational moments of the Galilean satellites of Jupiter and icy satellites of Saturn. *Icarus* 61 (1): 92–100.

Ziman, Ia. L., Krasikov, V. I., et al. 1975. Selenocentric systems of coordinates on the eastern sector of the far side of the Moon. In I. Efrevnov, ed., *Atlas of the Far Side of the Moon*, part 3, Razdel II. Moscow: Nauka, pp. 52–8 (in Russian).

6

Topographic mapping

SHERMAN S. C. WU AND
FREDERICK J. DOYLE

6.1. INTRODUCTION

Planetary topographic maps provide a quantitative representation of the landforms and relief of planetary surfaces. Such data lead to a better understanding of the geologic and tectonic histories of planets and their satellites. For example, topographic data are used to infer the properties, strength, and composition of planetary crusts and interiors, the nature of their gravitational fields, and the distribution of anomalies. Additionally, topographic data are essential for planning future missions to explore the solar system, particularly those involving spacecraft landings on planetary surfaces.

Topographic portrayal of bodies other than Earth and of satellites presents many unprecedented problems: (1) No oceans have been discovered, and thus sea level cannot be used as a reference surface for elevations; (2) geodetic and supplementary control networks are initially absent and, once established, are not as accurate or precise as those on Earth; (3) topographic data come from a variety of sources of different types, resolution, accuracy, and format; and (4) the surfaces and atmospheres of planetary bodies have photometric properties that are distinctly different from those of Earth. For example, the atmosphere of Venus is opaque to visible and near-visible wavelengths. Measurements of its surface must therefore be made by radar rather than on visible images, and a completely different data-reduction technique is required to produce topographic maps.

Although spacecraft have now visited twenty-five planets and satellites, information to support topographic mapping is available only for the Moon, Mars, Venus, and to a limited extent the Uranian satellite Miranda. Planimetric mapping, i.e., without topographic contours, has been conducted for all planets and satellites for which image data are available. This type of mapping is described in Chapter 3.

6.2. TOPOGRAPHIC DATA TYPES

For the Earth, topographic mapping at scales smaller than about 1:10,000 is performed almost exclusively with aerial photography, using the well-

Table 6.1. *Film cameras for lunar topographic mapping.*

Mission	Camera	Areas mapped	Focal length (mm)	Film format (mm)	Remarks
Zonds 6, 8		Single strips of lunar farside			USSR missions
Apollos 8 and 11–17	Hand-held Hasselblad from orbit	Various scientific sites	80–500	70 × 70	Uncalibrated
Apollos 13, 14	Hycon	Descartes	480	115 × 115	Uncalibrated
Apollos 15–17	Fairchild Metric	Apollo orbit tracks	75	115 × 115	Calibrated with reseau
Apollo 15–17	Itek Panoramic	Apollo orbit tracks	605	115 × 1120	High resolution
Apollo 11, 12, and 14–17	Hand-held Hasselblad from surface	Apollo landing sites	60	70 × 70	Calibrated with reseau

established techniques of stereophotogrammetry (American Society of Photogrammetry, 1980). With the single exception of the metric camera used on Apollo lunar missions 15, 16, and 17 (Table 6.1), the imaging systems on planetary exploration spacecraft have not been designed with topographic mapping as a primary objective. Use of film cameras implies that the film will be returned to Earth for processing and analysis. This has been possible only for lunar missions (Table 6.1).

For the planets and their satellites, imaging systems have been narrow-angle television cameras or electro-optical scanning systems. As a consequence, classical procedures have had to be extensively modified to adopt them to the geometric configuration of these sensors. The production of useful contour maps is more a tribute to the skill of the photogrammetrists involved than to the design of the imaging systems.

In addition to the imaging systems, various other devices have been employed to derive topographic information (Table 6.2). These do not provide stereodata from which topographic contours can be extracted directly; rather, they produce terrain section profiles or elevations of discrete points.

6.3. TOPOGRAPHIC DATUM

Mapping of the Earth and other planetary bodies requires two reference surfaces: (1) a mathematical figure on which positional coordinates, latitude and longitude (ϕ, λ), can be computed and (2) an equipotential surface from which topographic elevations can be measured.

Table 6.2. *Data sources for planetary topographic mapping.*

Instrument	Mission	Planetary body	Vertical accuracy	Data type	Reference
Goldstone radar	Earth-based	Mars	150 m	Profiles	Goldstein et al., 1970; Downs et al., 1971, 1973, 1975
Arecibo radar	Earth-based	Venus		Images	
Haystack radar	Earth-based	Mars	100 m	Profile	Pettengill et al., 1969, 1971; Rogers et al., 1970
Ultraviolet spectrometer	Mariner 9	Mars		Profiles	Barth and Hord, 1971
Infrared spectrometer	Mariner 9	Mars	1 km	Profiles	Herr et al., 1970; Hanel et al., 1970, 1972
Radio occultation	Mariner 9	Mars	0.3–2.1 km	Points	Kliore et al., 1972, 1973; Christensen, 1975
	Viking Orbiter	Mars		Points	Lindal et al., 1979
Laser altimeter	Apollos 15–17	Moon	2 m	Points	Doyle et al., 1977
Radar altimeter	Pioneer 10 Veneras 15, 16	Venus		Profiles	Pettengill et al., 1980; Barsukov et al., 1984
Stereophotogrammetry	Apollos 15–17	Moon	30 m	Contours	Schimerman, et al., 1973
	Mariner 9	Mars		Contours	
	Viking Orbiter	Mars	30–60 m	Contours	Wu et al., 1982
	Viking Lander	Mars	0.1–0.7 m	Contours	
	Voyager 2	Miranda			Wu et al., 1987

6.3.1. *Planetary reference figures*

According to the laws of physics, any elastic body that rotates about a body-fixed axis will assume the shape of an ellipsoid of revolution with a circular equator and elliptical meridians. The size and elasticity of a planetary body, its mass, its rotational velocity, the strength of the materials of which it is made, and its distance from the body about which it revolves – all have a profound effect on the shape of its elliptical surface. Thus, the giant planet Jupiter, composed mostly of low-density gases and rotating a full 360 degrees on its axis in about 10 hours, has elliptical meridians with eccentricities of nearly 0.354. Venus, on the other hand, is about the size of the Earth and is composed of strong, Earthlike materials, but it rotates very slowly. Only very sensitive instruments would be able to measure its ellipticity. The Moon is also composed of Earthlike materials and rotates very slowly; it has virtually no flattening at the poles. Several satellites of the outer planets are small bodies composed of heterogeneous mixtures of rock and ice. Their shapes may depart substantially from ellipsoids of revolution. Similarly, many satellites and asteroids are so small that their gravity is insufficient to form even approximately spherical shapes. Long periods of precise observations and extensive computation are required to determine an accurate figure. For some small bodies,

Table 6.3. *Reference surfaces for planetary bodies.*

Planetary body	Reference surface	Equatorial radius A (km)	Secondary radius B (km)	Polar radius C (km)
Earth	Biaxial	6,378.137		6,356.752
Moon	Spherical	1,736.0		
Mars	Biaxial	3,393.4		3,375.8
Phobos	Triaxial	13.5	11.5	9.5
Deimos	Triaxial	7.5	6.1	5.5
Jupiter	Biaxial			
Europa	Spherical	1,569		
Ganymede	Spherical	2,631		
Callisto	Spherical	2,400		
Io		Spherical	1,815	
Saturn				
Mimas	Biaxial	197	177–86	
Dione	Spherical	559		
Rhea	Spherical	764		
Enceladus	Biaxial	251	243	
Tethys	Spherical	524		
Iapetus	Spherical	718		
Venus	Spherical			
Mercury	Spherical	2,439		
Uranus				
Miranda	Triaxial	241,235	232	
Ariel	Spherical	579		
Umbriel	Spherical	586		
Titania	Spherical	790		
Oberon	Spherical	762		

a triaxial ellipsoid with an elliptical equator and meridians of varied ellipticity may actually provide a better fit. Triaxial ellipsoids have also been fitted to the Earth and other large bodies, but for the most part these are computational niceties with little practical value.

A unified reference ellipsoid for the Earth is described by the World Geodetic System 1984 (Defense Mapping Agency, 1987). The techniques for deriving planetary reference surfaces for other solar system bodies are described by Davies in Chapter 5. The currently accepted values are listed in Table 6.3. Their accuracy will undoubtedly improve as planetary exploration continues.

Before a reference surface can be used to produce graphic map products, it must be transformed to a plane surface through techniques of map projection, described by Batson in Chapter 3. Because conventional map projections are defined only for spherical or ellipsoidal bodies, simple reference surfaces are the most practical.

6.3.2. *Equipotential reference for elevations*

The compilation of terrestrial topographic maps is based, in part, on the assumption that water flows from higher locations down the slopes and valleys indicated by contour lines to a zero elevation at mean sea level. Mean sea level is an equipotential surface, i.e., one at which the value of gravity is a constant, and the direction of gravity is normal to the surface. This surface is called the *geoid*, and it is the reference for measuring topographic elevations. If there were no relief and the body were completely homogeneous, the geoid would coincide with the mathematical reference surface, and the cartographer's life would be easy – but dull. As it is, mountain heights, ocean abysses, and the uneven distribution of various materials composing the Earth (*mass anomalies*) cause the geoid to depart significantly from the ellipsoid. These departures are called the *undulations* of the geoid. On Earth they are determined by an extensive network of observations of the value of gravity (*gravity anomalies*) and its direction (*deflection of the vertical*).

The same physics applies to other Solar System bodies, but no extensive networks of gravity observations are available. However, the orbits of spacecraft are also controlled by gravity, and by careful and extended tracking of spacecraft positions, it is possible to derive the gravity field and hence the planetary geoid.

The spherical harmonic expansion used to represent gravity potential is

$$U = \frac{GM}{r} \left\{ 1 - \sum_{n=2}^{\infty} J_n \left(\frac{A}{r} \right)^n P_{no} \sin \phi \right. \tag{6.1}$$

$$+ \sum_{n=2}^{\infty} \sum_{m=1}^{n} P_{nm} \left(\frac{A}{r} \right)^n \sin \phi$$

$$\left[C_{nm} \cos m\lambda + S_{nm} \sin m\lambda \right] \Big\}$$

in which G is the constant of gravitation, M is the mass, r is the radial distance of the orbiter from the planetary body, A is the equatorial radius, ϕ is the latitude, P_{nm} denotes the associated Legendre polynomial, λ is the longitude, and C_{nm}'s and S_{nm}'s are the gravity coefficients of the planetary bodies to be considered.

If $-J_n = C_{no}$, $\mu = GM$, ω = rotational velocity of the body, and R_m = an arbitrarily selected radius of the mean sphere, then the geopotential for ΔR_m is

$$\Delta R_m = \Delta r + \frac{1}{2} \omega^2 \frac{R_m^4}{\mu} \cos^2 \phi \tag{6.2}$$

In this expression

$$\Delta r = R_m \sum_{n=1}^{N} \sum_{m=0}^{n} P_{nm} (\sin \phi) \left[C_{nm} \cos m\lambda + S_{nm} \sin m\lambda \right] \tag{6.3}$$

is the spherical harmonic expansion used to express a planetary gravity field in terms of radial deviation from a mean sphere (Jordan and Lorell, 1975). The topographic datum of a planetary body then can be defined by R_o:

$$R_o = R_m + \Delta R_m \tag{6.4}$$

There is little point in using the geoid in preference to the ellipsoid unless the precision of elevation measurement is at least the same order of magnitude

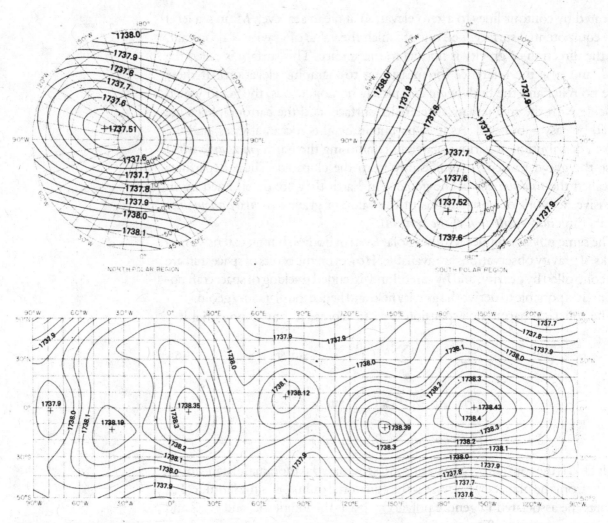

NORTH POLAR REGION

SOUTH POLAR REGION

Figure 6.1
Contour map of the topographic datum of the Moon. This figure is derived from the lunar gravity field with a mean radius of 1738 km. Contour interval is 50 m.

as the geoid undulations. In addition, from a practical point of view, liquid water has yet to be discovered on any other planet. Nevertheless, the geoid has practical significance for mass wasting, geophysical interpretations, and vehicle movement across the surface.

6.3.3. Equipotential surface for the Moon

Until 1981, all lunar topographic maps were compiled on a spherical datum. An equipotential reference surface for the Moon was computed by Wu (1981a, 1985); the datum is based on the lunar gravity field expressed in terms of spherical harmonics of fifth degree and fifth order, with the sixth-degree sectorial terms. Gravity coefficients C_{nm} and S_{nm} used for the derivation of the datum were developed by Sjogren (Ferrari et al., 1980), who worked with Lunar Orbiter 4 tracking data and laser ranging data. The datum was computed by using $R_m = 1738$ km as the mean radius and by equation 6.4, where

Table 6.4. *Topographic datum of the Moon. Values of the radius (in kilometers) at selected coordinates of latitude and longitude.*

Latitude	Longitude						
	180°W	120°W	60°W	0°	60°E	120°E	180°E
90°N			1737.550				
60°N	1737.849	1737.591	1737.778	1737.971	1737.771	1737.796	1737.849
30°N	1738.098	1737.990	1737.067	1738.224	1737.964	1737.965	1738.098
0°	1738.309	1738.252	1738.032	1738.270	1738.054	1738.179	1738.309
30°S	1738.019	1738.042	1738.080	1738.098	1737.986	1738.055	1738.019
60°S	1737.544	1737.709	1737.880	1737.795	1737.901	1737.800	1737.544
90°S			1737.748				

$\omega = 0.26617 \times 10^{-5}$ rad/sec and $\mu = 4902.799$ km^3 sec^{-3}. It is graphically shown as the contour map in Figure 6.1. The values of R_o for a 5-degree increment in both longitude and latitude were computed (Wu, 1981a) and are also listed (in larger increments in longitude and latitude) in Table 6.4.

6.3.4. Equipotential surface for Mars

Mars completes a rotation in approximately 24 hours, producing significant polar flattening that is measurable through Earth-based telescopes. The radio signal from the Mariner 9 spacecraft was occulted by Mars twice in every 24-hour period, thus providing two radius measurements per day. A total 256 measurements were made during the entire mission, scattered over the surface between latitude 86° N and latitude 80° S. The uncertainty of these points was calculated to be between 0.33 km and 1.1 km, with a worst-case error of 2.1 km (Kliore et al., 1973; Christensen, 1975). The measurements show that the average terrain elevation in the northern hemisphere is 3 to 4 km lower than in the southern hemisphere.

Irregularities in the gravity field of Mars have been mapped by measuring accelerations and decelerations of orbiting spacecraft (Lorell et al., 1970, 1972). A rough gravity field requires large coefficients in the spherical harmonic representation. However, stability occurs only at fourth-degree and fourth-order spherical harmonics (Jordan and Lorell, 1975); the fourth-degree and -order were therefore adopted for computing the Mars topographic datum. Values of the gravity coefficients C_{nm} and S_{nm} used for computing the datum were obtained by Jordan and Lorell (1975) from the results of the celestial mechanics experiment of Mariner 9. The values of R_o for a 5-degree increment in both longitude and latitude were computed (Wu, 1981a). They are listed in larger increments in Table 6.5.

A triaxial ellipsoid figure can be used to approximate this Mars topographic datum. Parameters of the geometric figure are $A = 3394.6$ km, $B = 3393.3$ km, $C = 3376.3$ km, and $\theta = $ longitude 105°. This topographic datum is

Table 6.5. *Topographic datum of Mars: Values of the radius (in kilometers) at selected coordinates of latitude and longitude.*

Longitude	0° = 360°	60°	120°	180°	240°	300°
Latitude						
90° N	3,376.249					
60° N	3,380.530	3,380.527	3,380.646	3,380.333	3,380.404	3,380.952
30° N	3,389.197	3,389.323	3,390.252	3,388.807	3,389.765	3,389.879
0°	3,393.633	3,393.693	3,395.009	3,393.195	3,393.976	3,394.269
30° S	3,389.237	3,389.279	3,389.844	3,388.972	3,389.558	3,389.694
60° S	3,380.661	3,380.630	3,380.634	3,380.642	3,380.937	3,380.902
90° S	3,376.492					

shown by contours with respect to a mean sphere with a radius of 3382.92 km in Figure 6.2.

A unique procedure was employed to determine Mars' zero elevation. The Ultraviolet Spectrometer (UVS) on Mariner 9 was designed to derive vertical temperature profiles in the Martian atmosphere by measuring ultraviolet reflections. Because the surface of Mars does not reflect ultraviolet radiation, all reflections recorded by the instrument must come from the Martian atmosphere, and thus their intensity is inferred to represent atmospheric pressure. Barth and Hord (1971) measured the atmospheric pressure of nearly seventy-five hundred points along thirty-nine suborbital tracks between latitude 60° S and 45° N. The zero elevation is defined as lying where the mean atmospheric pressure is equal to the triple point of water (6.1 millibars, the pressure and temperature at which water could theoretically exist simultaneously as a gas, a liquid, and a solid).

6.3.5. Vertical datum for Venus

The gravity data from the Pioneer Venus mission indicate that the difference between Venus' radii at the equator and at the two poles is not larger than 50 m (Sjogren et al., 1981). A spherical figure with a radius of 6051.5 km is therefore adequate to represent the elevation reference system.

6.4. SECONDARY CONTROL NETWORKS

Before extensive topographic mapping can be conducted, a network of secondary control points must be established, for which latitude, longitude, and elevation have been computed with respect to the selected reference systems. The most efficient method for developing secondary control points is through the process of aerotriangulation. This technique has been perfected for terrestrial mapping by wide-angle photogrammetric cameras carried in aircraft (Slama et al., 1980), and it has been modified to make it adaptable to lunar and planetary mapping.

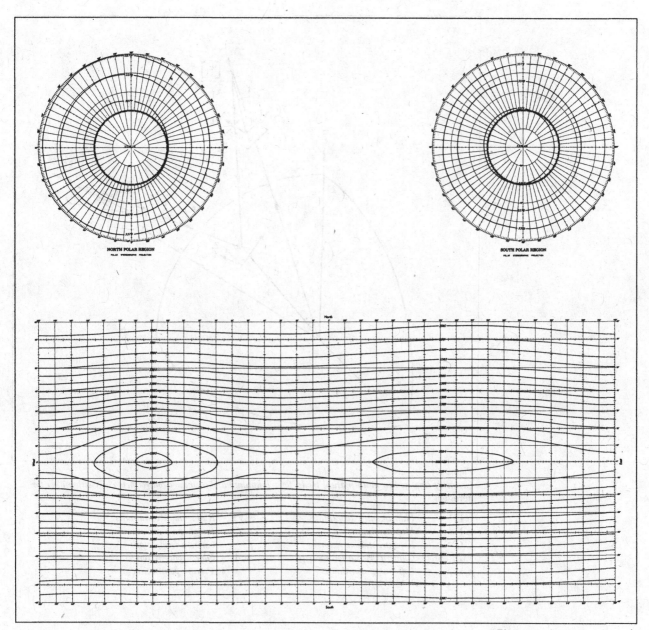

Figure 6.2

Contour map of the topographic datum of Mars. This figure is derived from the Martian gravity field with respect to a sphere with a mean radius of 3382.92 km. Contour interval is 1 km.

6.4.1. *Planetary triangulation*

The basic geometry of triangulation is illustrated in Figure 6.3. The location of a terrain point P is given in the planetary coordinate system XYZ. Each terrain point must be imaged on at least two separate photographs taken from exposure stations C_1 and C_2, located in the same planetary coordinate system. Each photograph has its own coordinate system, $x_1 y_1 z_1$ and $x_2 y_2 z_2$, defined by reference marks in the image plane and the focal length of the optical system. The images p_1 and p_2 of the terrain point can be measured in each image

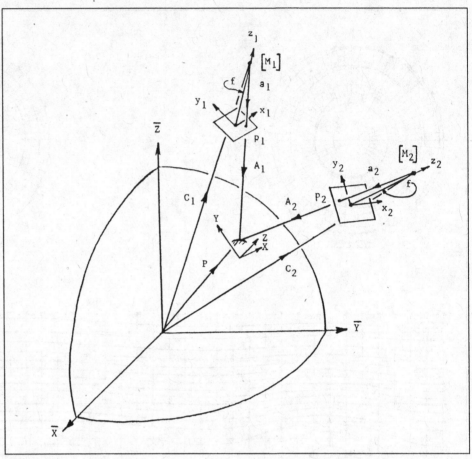

Figure 6.3
Geometry employed in mathematical model for triangulation of secondary control nets. See Section 6.4.1 for discussion.

$\overline{XYZ}$	*Planetocentric coordinates*	$x_2 y_2 z_2$	*Right image coordinates*
XYZ	*Map coordinates*	a, a_2	*Image space vectors*
C, C_2	*Exposure station vectors*	f	*Camera focal length*
$[M_1] [M_2]$	*Camera orientation matrices*	A, A_2	*Object space vectors*
$x_1 y_1 z_1$	*Left image coordinates*	P	*Terrain point vector*

coordinate system to define vectors a_1 and a_2, looking at the terrain point. The orientation of each image coordinate system in the planetary coordinate system is defined by three angles, ωφκ, which make up an orientation matrix [M]. When an image vector a is multiplied by this orientation matrix, it is transformed into a corresponding vector A in the planetary coordinate system. The solution for the intersection of two (or more) such vectors gives the location of the terrain point P. In implementing the computations, large numbers of points and large numbers of photographs are involved in a least-squares solution that adjusts the locations C and orientations [M] of the photographs and computes the XYZ locations of all terrain points.

The fundamental photogrammetric equations that relate these quantities are

$$x_p = \frac{-f[m_{11}(X_p - X_c) + m_{12}(Y_p - Y_c) + m_{13}(Z_p - Z_c)]}{[m_{31}(X_p - X_c) + m_{32}(Y_p - Y_c) + m_{33}(Z_p - Z_c)]} \quad (6.5)$$

and

$$y_p = \frac{-f[m_{21}(X_p - X_c) + m_{22}(Y_p - Y_c) + m_{33}(Z_p - Z_c)]}{[m_{31}(X_p - X_c) + m_{32}(Y_p - Y_c) + m_{33}(Z_p - Z_c)]} \quad (6.6)$$

In these equations, x_p and y_p = measured coordinates of image point p; f = focal length of optical system; m_{ij} = elements of orthonormal matrix [M], which is composed of the three rotation angles, $\omega\phi\kappa$, defining the orientation of the camera coordinate system in the planetary coordinate system; X_p, Y_p, and Z_p = planetary coordinates of terrain point P; and X_c, Y_c, and Z_c = planetary coordinates of camera exposure station C.

These two equations can be written for each image point that appears on a photograph. A massive least-squares solution adjusts the values of exposure-station positions and orientations and ground-point coordinates by minimizing the residuals in the observed image coordinates.

Traditional photogrammetric triangulation relies on the wide fields-of-view of aerial cameras, which result in triangles with apex angles that are considerably larger than the errors involved in measuring those angles. The photogrammetric block then becomes a rigid geometric structure. Not only the locations of intersecting vectors to ground points, but also the locations and orientations of the camera perspective centers can be precisely determined. Only the scale and orientation of the block are not implicit in its internal geometry, requiring a minimum of three external control points, provided by field surveys, to define the block fully.

The fields-of-view of most cameras used in planetary mapping are so narrow that they cannot provide the necessary rigidity to the photogrammetric block. Furthermore, external control points are not usually available to provide orientation and scale. With sufficiently accurate knowledge of camera parameters (position and orientation angles), however, the internal geometry of the block can be defined, as well as its scale and orientation, even without external control. Fortunately, spacecraft navigation by radio-tracking of signals from spacecraft has become sufficiently precise to allow accurate computation of the spacecraft positions with respect to their targets, thereby making mapping possible. Special techniques for dealing with these conditions were discussed by Wu et al. (1982). Occasionally, photogrammetric integrity can be further enhanced by precise altimetry provided by laser-ranging instruments on spacecraft and by star cameras (mounted on the same platform as the survey cameras) that provide accurate information on platform orientation.

Planetocentric coordinates $\overline{X}\,\overline{Y}\,\overline{Z}$ are not directly useful for mapping. They are first transformed into geodetic coordinates, latitude, longitude, and ele-

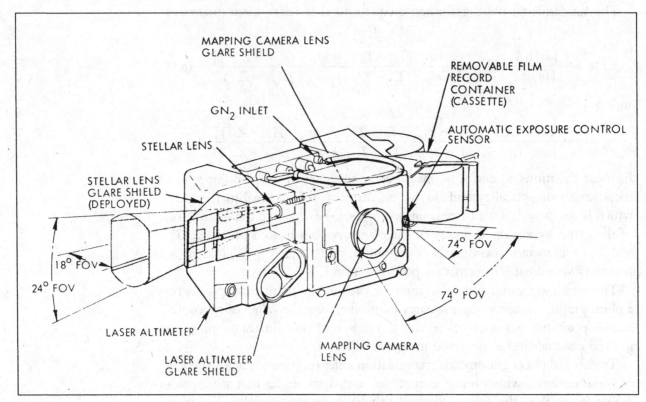

Figure 6.4
Metric camera system carried on Apollos 15, 16, and 17. See Section 6.4.2 for discussion.

vation by well-known geodetic formulas; they are then converted to local coordinate systems *XYZ* by using the map projections described in Chapter 3.

6.4.2. *Lunar control net*

The metric camera system carried on Apollos 15, 16, and 17 was the only sensor configuration defined explicitly for topographic mapping (Figure 6.4). The system consisted of the calibrated mapping camera with 75-mm focal length and 115- × 115-mm film format. Boresighted with the camera was a laser altimeter that measured the distance from the spacecraft to the lunar surface each time a photograph was taken. Rigidly attached to the mapping camera was a stellar camera that also photographed the star field each time a photograph was taken. Measurement of the stellar images permitted accurate determination of the attitude angles of the mapping camera.

These cameras took 1250 photographs in twenty-four orbital paths. The photographs have a nominal spatial resolution of 25 m, and the entire collection covers approximately 20 percent of the lunar surface (Figure 6.5). Apollos 15, 16, and 17 also carried a panoramic camera of 605-mm focal length operating with 115-mm film. It was designed essentially to provide high-resolution data, and the best of the photographs have about 2-m ground resolution from 110-km orbit. The geometric integrity of these pictures was

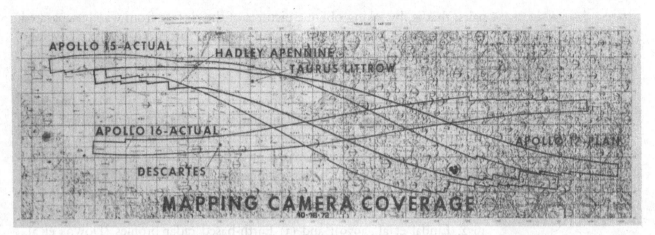

Figure 6.5
*Coverage by metric camera on
Apollos 15, 16, and 17.*

very poor, and they were used for mapping by identifying the secondary control points computed from the metric camera triangulation.

Lunar topographic triangulations were performed with the Apollo dataset by two different agencies. The Defense Mapping Agency Aerospace Center (DMAAC) (Schimerman, 1973) used the orbital tracking data to constrain computations made with Apollo 15 photographs. They later tied the Apollo 16 and 17 nets to the Apollo 15 net. The National Oceanic and Atmospheric Administration (NOAA) and the U.S. Geological Survey (USGS) (Doyle et al., 1977) performed a purely photogrammetric solution that obtained an adjustment of 23,436 unknowns (coordinates of control points and parameters of camera stations) by simultaneously using the data of all three missions. The laser altimeter observations were used as a scale restraint in each stereomodel, and the attitudes of each photograph were constrained by the values obtained from the stellar camera reductions. In the NOAA/USGS net, 70 percent of the adjusted control points have a standard error of less than 30 m, and 74 percent of the elevations have a standard error of less than 30 m. This control network served as the basis for the most precise topographic maps of the Moon.

6.4.3. *Mars control net*

The orbiting vehicles of the Mariner and Viking Missions carried narrow-angle television cameras, which enabled them to obtain high-resolution images. The data that they recorded was digitized, tape recorded, and later transmitted to Earth. However, television images are difficult to use for photogrammetric mapping. Image formats are small: typically 800 lines by 800 samples of image data, on an 11- × 11-mm vidicon image plane. Electronic transmission of images introduces severe geometric distortions that must be corrected digitally before the images can be used for mapping. Stereoscopy can be achieved with narrow-angle television images by taking convergent pairs, that is, by taking oblique pictures of the surface from widely separated points in space. Most stereopairs have been the fortuitous result of coverage sequence planning unrelated to photogrammetric requirements for mapping.

PRELIMINARY NETWORK   A topographic control net for Mars has been compiled by photogrammetric triangulation and tied to existing computations of horizontal and vertical positions of features on the planet. These computations include (1) the planimetric control net containing more than six thousand points on Mars, which was compiled from Mariner 9 and Viking Orbiter pictures (Davies, 1973; Davies et al., 1978); (2) occultation measurements based on the precise moment a spacecraft passed behind Mars (Kliore et al., 1972; Lindal et al., 1979); and (3) Earth-based radar profiles (Downs et al., 1975). Davies imposed an independently derived radius at each point; computation of the radius was not part of the derivation of the net itself. The net is therefore adequate for planimetric mapping but not for topography. Occultation observations provide precise radius measurements, but the exact longitude of the point that caused the occultation (i.e., a mountain or ridge on the Martian horizon as viewed from Earth) cannot be determined directly. The location on Mars of Earth-based radar profiles is measured in terms of celestial mechanics rather than on photogrammetrically triangulated positions of surface features. Although the profiles have very high vertical precision, it would be fortuitous indeed if horizontal agreement with existing planimetric controls were precise enough to control topographic mapping.

REVISED NETWORK   A special stereoscopic survey of Mars was taken by the Viking Orbiter spacecraft late in the mission. This survey resulted in a set of 1,180 images that provide stereoscopic coverage of the Martian surface. The nominal resolution of these images is 750 to 1,000 m per pixel; they were taken from altitudes as high as 37,000 km above the Martian surface.

The topographic control network was compiled by using the USGS General Integrated Analytical Triangulation (GIANT) program (Elassal, 1976; Elassal and Malhotra, 1989), which fundamentally involves the least-squares solution of large numbers of equations 6.5 and 6.6. This program computes the geographic coordinates of selected control points. It requires that the user identify images of these control points on the pictures to be used in the survey and measure and record their precise locations in each picture frame. The control points must include those that are newly defined as well as existing controls that have been computed previously. The user may specify the weighting (i.e., the estimated error) to be given to the horizontal and vertical components of the existing controls. The program also requires estimates of the locations and orientations of the camera stations at the time each picture was taken, as well as the allowable deviation from those locations and orientations. Final output from the program includes geographic or geocentric coordinates of all control points, with previously computed points shifted within allowable limits, as well as recomputed camera station locations and camera orientation angles within limitations imposed by the user.

The camera locations were constrained to lie within 2 km of locations predicted by radio-tracking data. Viking camera orientations are known within 0.25 degrees, based on information telemetered from the spacecraft. This pointing constraint was also imposed on GIANT. In order to tie the topographic net as closely as possible to the existing planimetric net, the planimetric locations of the Davies control points were given weights of ±15 minutes, compared with weights of ±1 degree given to the newly computed points. The horizontal positions of radar and occultation points were given weights of only ±2 degrees, whereas their vertical weights were 500 m.

The control net was originally compiled in two bands. The equatorial band extends from latitude 30° S to 30° N. A second band, centered on the 120°/ 300° meridian, covers approximately 30 degrees. A total of 3,172 topographic control points were produced by the first effort. Residual errors in the adjustment are approximately 4 km horizontally and 750 m vertically.

A supplemental net, containing 465 Viking Orbiter images, was computed to fill in the 30 percent of the planet not covered by the original two bands. The stereoscopic geometry in these areas is poor, resulting in residual errors of approximately 5 km horizontally and 800 m vertically.

6.5. MAP COMPILATION TECHNIQUES

The wide variety of data sources for planetary topographic mapping, listed in Tables 6.1 and 6.2, has required the modification of conventional mapping techniques and the development of completely new and innovative procedures.

6.5.1. *Stereoscopic mapping*

Topographic contour maps are commonly made by stereophotogrammetry, a process whereby overlapping images are viewed stereoscopically on specially designed machines called stereoplotters. There is a wide variety of such instruments, primarily designed for use with conventional aerial photographs of the Earth. Because of the peculiar geometries of planetary imaging systems, the type of instrument most commonly used is the analytical stereoplotter (Figure 6.6). The optical system of the stereoplotter allows an operator to view a pair of stereoscopic images simultaneously; a measuring mark appears to float in the stereoscopic field, or model. The apparent position of the mark can be moved very precisely in three dimensions by the operator. This position is coupled mechanically or electronically, depending on the plotter design, with a stylus on a plotting table. The operator traces contour lines by setting the vertical position of the mark at some defined value and moving it horizontally until it appears to touch a landform. He then lowers the stylus onto the drawing medium and moves the mark about a horizon-

Figure 6.6
Analytical stereoplotter AS-11AM used by the U.S. Geological Survey to compile topographic contours for planetary mapping.

tal plane in the model, keeping it always in contact with the stereoscopic image of the surface being mapped; the stylus simultaneously traces this contour line.

The function of an analytical stereoplotter (Figure 6.7) is to solve continuously the projective equations 6.5 and 6.6 for two photographs simultaneously. The constant values – focal length f, exposure station positions C_1 and C_2, and orientation matrices $[M_1]$ and $[M_2]$ – are input to the control computer. By looking through the instrument's binocular optical system at photograph 1 with the left eye and photograph 2 with the right eye, the operator

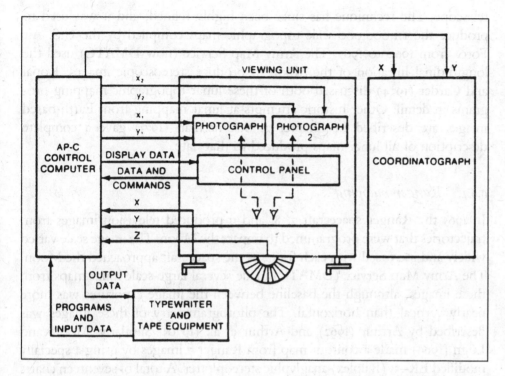

Figure 6.7
Schematic design of analytical
stereoplotter. See Section 6.5.1 for
explanation.

sees a stereomodel of the terrain that includes the floating mark. By turning
the handwheels, the operator can move the mark horizontally to XY coordinates of the terrain point P and, by turning the footwheel, adjust the elevation coordinate Z. The control computer calculates the image coordinates
x_1y_1 and x_2y_2 and drives the photo stages until the corresponding image points
are seen by the operator. The horizontal translations are directed to the coor-
dinatograph, which draws the contour line while its elevation is recorded by
the operator. The planetary coordinates can also be recorded digitally as the
map is compiled.

6.5.2. *Mapping with other data types*

Other data types, such as Earth-based radar, laser and radar altimetry, and
spectroscopic sensors, can contribute to topographic mapping. For the most
part these data types are unique to each mission, and no standard procedures
have been developed for their use. Consequently, they are described in conjunction with the specific planetary mapping projects on which they were
employed.

6.6. LUNAR TOPOGRAPHIC MAPPING

Herschel, in 1787, made the earliest topographic measurements of the Moon
by measuring shadows of mountain peaks with a micrometer mounted on his

telescope. The technique has since been highly refined, and it was used to produce the 1:1,000,000-scale topographic maps compiled by the U.S. Air Force from 1960 to 1967. The Army Map Service (now DMATC) used the longitudinal libration of the Moon to produce stereoscopic images. Kopal and Carder (1974) discussed both of these lunar topographic mapping programs in detail. Other historic attempts at lunar mapping from Earth-based images are described in Chapter 2. Schimerman (1973) gave a complete description of all lunar maps produced to that date.

6.6.1. Ranger mapping

In 1965 the Ranger spacecraft 7, 8, and 9 produced television images from trajectories that were programmed to impact the Moon. The image scale varied widely and increased with each frame as the spacecraft approached the Moon. The Army Map Service (DMATC) made several large-scale lunar maps from these images, although the baseline between the image exposures was more nearly vertical than horizontal. The photogrammetry of these images was described by Arthur (1962) and Arthur et al. (1970). Similarly, Moore and Lugn (1966) made a contour map from Ranger 7 images by using a specially modified ER–55 (Balplex) anaglyphic stereoplotter. A total of seventeen charts were compiled for potential Apollo landing sites, at scales ranging from 1:1,000,000 to 1:1,000.

6.6.2. Lunar orbiter mapping

Lunar Orbiters 1 to 5 photographed nearly 99 percent of the Moon's surface. The photographic system included a dual-lens camera, an on-board film-processing subsystem, an electronic film-scanning and communication subsystem, and a ground reconstruction subsystem. In Figure 6.8, the diagram at the left shows the photographic subsystem, which projected a wide-angle picture, using an 80-mm lens, and a high-resolution telescopic picture, using a 610-mm lens, onto a single 70-mm film web. The film was then developed onboard and transported to the readout scanner, shown at the right of Figure 6.8. Here, the film was electronically scanned in 2.54-mm increments across the width of the film, and the resulting video signal was transmitted to Earth. The ground reconstruction subsystem converted the signal to an intensity-modulated line on a cathode ray tube. In a continuous-motion camera with 7.2× enlargement, 35-mm film was pulled past the image of this line, producing a series of 35-mm framelets that could be reassembled into enlarged replicas of the original spacecraft frames. This ingenious system worked very well, although most of the resulting pictures show conspicuous lines where the framelets join, and it was extremely difficult to maintain photogrammetric integrity for mapping.

The primary cartographic use of Lunar Orbiter photography was to compile small-scale planimetric maps of the nearside and farside of the Moon, with

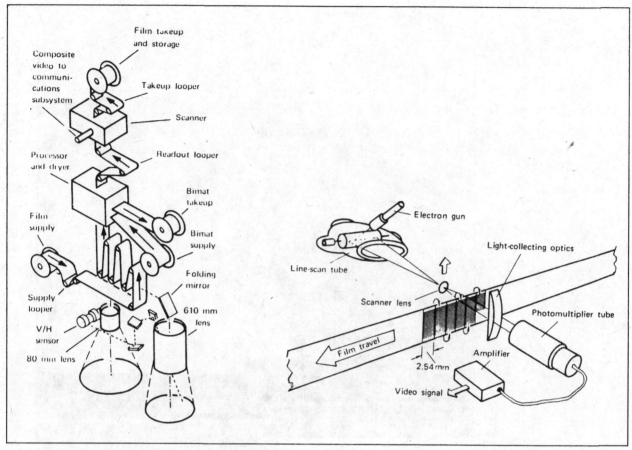

Figure 6.8
The photographic subsystem of Lunar Orbiter. The dual-lens camera system is at the left, and the scanning and transmission system is at the right.

shaded relief to indicate topography. The maps provide detailed coverage of equatorial sites where the Apollo landing sites were eventually located. Topographic maps of these sites were also produced, but due to lack of control and poor geometric integrity of the reconstructed photographs, the mapping was purely relative and was not related to any horizontal or vertical datum.

During 1969–71, 1:250,000-scale topographic photomaps were produced for a number of sites of scientific interest. The map bases are green and black duotone prints. Maps based on Orbiter III photographs have 200-m contour intervals; those compiled from higher-altitude Orbiter V coverage have 400-m contour intervals. Figure 6.9 is a sardonic comment on the difficulty of topographic mapping with Lunar Orbiter photographs.

In 1968–70 the Soviet Union photographed the farside of the Moon with film cameras carried on the Zond 6 and 8 spacecraft and transmitted the images to Earth by a similar process. These electronically scanned pictures, while containing excellent high-resolution-image information, were not very useful for extensive photogrammetric mapping.

Well....I'll be damned !

Figure 6.9
The Moon's topography as recorded by Lunar Orbiter
photographs.

6.6.3. Apollo mapping

The Apollo Lunar program provided a variety of photographic sources used for topographic mapping.

HAND-HELD PHOTOGRAPHY   Many pictures, intended primarily for mission documentation, were taken with hand-held cameras from the orbiting Apollo command modules. The cameras had interchangeable lenses of 60-, 80-, 250-, and 500-mm focal length and were not calibrated for photogrammetric work. The images were usable, however, for mapping selected areas on local reference datums (Wu, 1969). Photogrammetrically calibrated hand-held cameras, primarily of 60-mm focal length, were used by the astronauts on the lunar surface for scientific documentation of the geologic traverses, and many very large-scale sample environment maps were later made from photographs taken by these cameras.

LASER ALTIMETRY   The laser altimeter attached to the metric camera system used on Apollo missions 15, 16, and 17 had a ground footprint of approximately 20 m and a measurement precision of about 1 m. The laser was used not only in conjunction with the camera but also to obtain continuous profiles completely around the Moon. Figure 6.10, which shows data acquired by Apollo 17, is representative. Harmonic analysis of these data from all three missions was combined with precise range measurements made with lasers between observatories on Earth and corner reflectors placed on the Moon by the Apollo 11, 12, and 14 astronauts and by the Soviet Union's Luna 17 and 21 spacecraft. Prior to Apollo 15, control compilations for the Moon were based on the assumption of a smooth sphere with a radius of 1,738.64 km (Meyer and Ruffin, 1965; Arthur, 1968.). These new analyses indicated that the Moon is spherical within about 4 km, that it is asymmetric in shape, and that it is slightly egg shaped, with the small end pointing toward Earth (French, 1977). The center-of-mass of the Moon is displaced toward Earth about 2 km from its center-of-figure (Bill and Ferrari, 1977).

In principle, the elevations determined by the altimeter could have served as vertical control points for stereoscopic contour mapping with the metric camera photographs, because the altimeter was boresighted with the camera. However, this procedure was not followed because the mapping and orbital analyses were not accomplished in the same time frame. Existing mapping is therefore based on photogrammetry alone.

HYCON CAMERA MAPPING   On Apollos 13 and 14 a Hycon reconnaissance camera, identified as the Lunar Topographic Camera (LTC), with 480-mm focal length and 115- × 115-mm format, was carried in the command module. This was considered an expedient until the metric camera system used on Apollos 15, 16, and 17 could become operational. Apollo 13 suffered an explosion in the service module en route to the Moon, and no photographs were taken with the LTC. On Apollo 14 a single strip of photographs, extending from Theophilus crater to the Kant plateau, was exposed. An electrical malfunction in the camera limited the mission to about two hundred useful photographs, which ended just short of the candidate landing site in the Descartes region. Though the photographs had a ground resolution of about 2 m, no significant topographic mapping was performed because the area of primary interest was not recorded.

METRIC CAMERA MAPPING   The metric camera photographs obtained from Apollos 15, 16, and 17 are the only truly photogrammetric material produced by the lunar and planetary exploration missions. The Apollo program was originally planned for a total of twenty missions. At least two of the remaining missions could have had polar orbiters, which would have resulted in nearly complete coverage. The failure to achieve complete coverage is one of the major disappointments resulting from the cancellation of the last three Apollo missions.

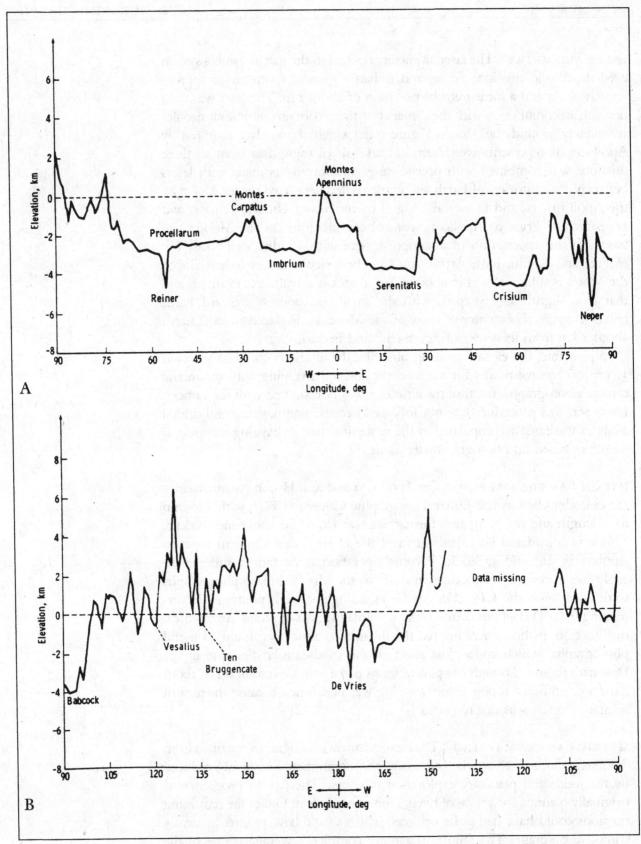

Figure 6.10
Laser altimeter traces acquired by Apollo 17 referred to a sphere with a radius of 1,738 km.

Topographic mapping was conducted by the Defense Mapping Agency Aeronautical Chart and Information Center (DMA-ACIC), now known as the Aerospace Center (DMAAC), and by the USGS at Flagstaff, Arizona. As mentioned in Section 6.4.2, two different secondary control networks were derived, and each organization used them to produce different topographic map products.

ACIC undertook a program to compile a Lunar Map (LM) Series to provide continuous topographic coverage at 1:1,000,000 scale. It was begun in 1976 as a replacement for the Lunar Astronautical Charts (LACs) at the same scale that had been compiled from Earth-based observations during 1962–7. The LM maps portray lunar topography by shaded relief in green as viewed with an eastern illumination. Spot elevations, crater depths, and a basic 300-m contour interval were also employed to provide relief information. In areas where Apollo coverage was not available, Lunar Orbiter photographs were used with a consequent decrease in accuracy. Only eleven sheets of the LM Series were published.

ACIC also undertook two series of maps at 1:250,000 scale. The Lunar Topographic Orthophotomap (LTO) and Lunar Orthophotomap (LO) Series are the most comprehensive and detailed maps resulting from Apollo mapping photographs, and they provide nearly continuous coverage within the area covered by these photographs. These series are also the results of the first major application of recent developments in orthophotography to lunar mapping. The complete photographic detail appearing in these orthophotomap bases is free of the positional displacements associated with the perspective view presented by normal photographs. The LTO editions display 100-m contours, 50-m supplemental contours, and spot elevations in a red overprint to the orthphotobase, which is lithographed in black and white. LO editions are identical except that all relief information is omitted and the selenographic graticule is restricted to border ticks, presenting an unencumbered view of lunar features imaged by the photographic base.

Metric camera mapping at the USGS Flagstaff facility consisted primarily of experimental maps and specific site maps produced to support geologic investigations. The NOAA/USGS supplementary control network, described in Section 6.4.2, was converted to the lunar datum, described in Section 6.3.3, and used to compile an experimental topographic map. This map shows that part of the Moon covered by the Apollo photographs (Wu 1981b, 1985); part of the map is shown in Figure 6.11.

An example of the larger-scale maps compiled from the metric camera photographs is shown in Figure 6.12A, of the crater Tsiolkovsky and vicinity. Profiles are often drawn from contour maps for the support of geologic studies, as shown in Figure 6.12B. Most of these detailed maps were compiled before completion of the secondary control net. Hence, their location is derived primarily from the spacecraft tracking data, and the elevations are based on a local datum. Standard errors of horizontal and vertical repeatability from the metric photography on the analytical stereoplotter are ± 7.4 m and ± 16.4 m,

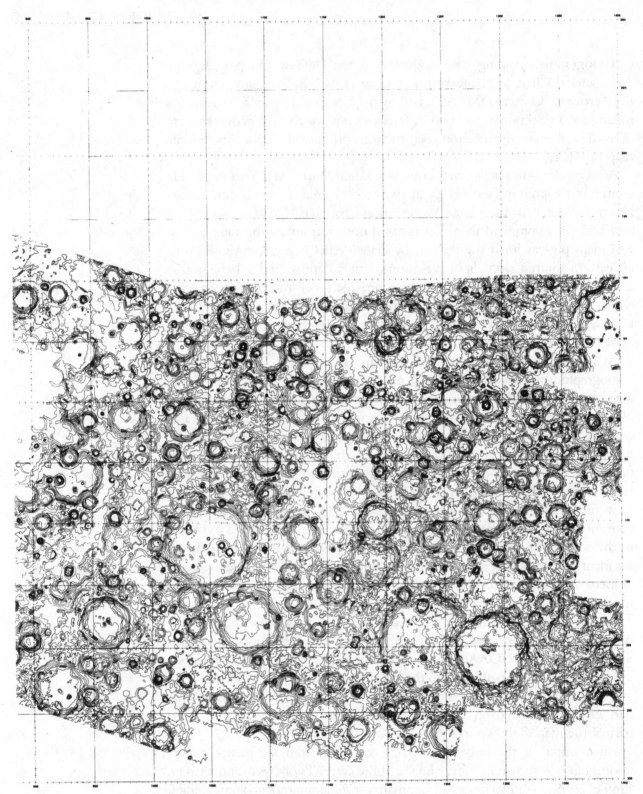

Figure 6.11
Part of an experimental contour map of the Moon at 1:2,750,000 scale with 500-m contour intervals, compiled from Apollo metric mapping camera photographs.

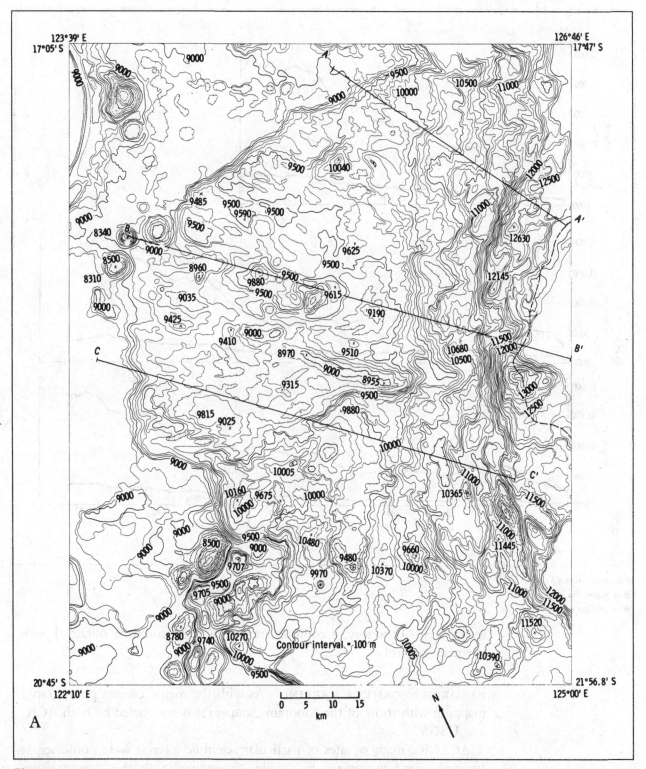

Figure 6.12

A. Topographic map of crater Tsiolkovsky compiled with 200-m contours from Apollo metric camera photographs.
(Continued)

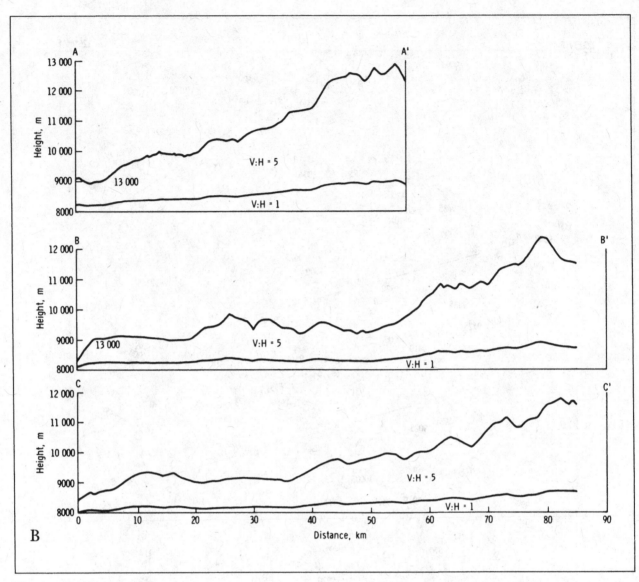

Figure 6.12 (cont.)
B. Topographic profiles drawn
from contour map of 6.12A.

respectively. Therefore, a contour interval of 50 m could be obtained with adequate control.

PANORAMIC CAMERA MAPPING   As with the metric camera photograph, mapping with those of the panoramic camera was conducted by both ACIC and USGS.

At ACIC, maps of sites of particular scientific interest were published at 1:10,000, 1:25,000, or 1:50,000 scale, depending upon the characteristics of the site being portrayed and the quality and applicability of specific photographic coverage to large-scale map compilation. The maps show spot elevations, 20-m contours, and 10-m supplements at 1:50,000 and

1:25,000 scales and 10-m contours and 5-m supplements at 1:10,000 scale. Relief information is carried on a red overprint to the photographic base, which is lithographed in black and white. Horizontal positions were related to the ACIC supplementary control network. Figure 6.13 shows one of these maps used as a base for a geologic map at the Apollo 15 landing site.

Representative of the work accomplished by the USGS with the panoramic camera is Figure 6.14, which shows the Davy crater chain. The map was compiled from photographs acquired by Apollo 16.

6.7. TOPOGRAPHIC MAPPING OF MARS

As with the Moon, topographic mapping of Mars has proceeded in several stages, depending upon the availability of data.

6.7.1.   *Mapping with Earth-based radar and radio occultation data*

Elevation profiles along the trace of the sub-Earth point on a rotating planet can be compiled by measuring the delay between the time a radar signal is transmitted from an observatory on Earth and the time the reflected signal is returned. Variations in the measured ranges are corrected for the relative motions of Earth and planet, and the residuals are interpreted as an elevation profile on the planet's surface. Spot sizes can be as small as 8 km, and elevation precision ranges from 75 to 200 m. The most accurate profiles are obtained from the equatorial regions of rapidly rotating planets. Such observations have been made of Mars by the Goldstone radar in California (Goldstein et al., 1970; Downs et al., 1971, 1973, and 1975) and by the Arecibo radar in Puerto Rico.

When a spacecraft passes behind a planet viewed from Earth, its radio signal is occulted. If the time of that occultation can be precisely measured, if the atmospheric composition and density are precisely known, and if the precise latitude and longitude of the occultation point are known, an accurate measurement of the radius of the planet at that point can be computed. Two of these conditions are usually met. The location of the point is less certain, however, because of the high probability that the signal was occulted by a mountain or ridge beyond the nominal horizon. (Christensen, 1975).

These two data types were incorporated in the derivation of the Martian reference surface described in section 6.3.4.

6.7.2.   *Mapping with atmospheric sounder data*

In 1971 the Mariner 9 spacecraft orbited Mars and carried the UVS and the Infrared Interferometer Spectrometer (IRIS), which were designed for atmo-

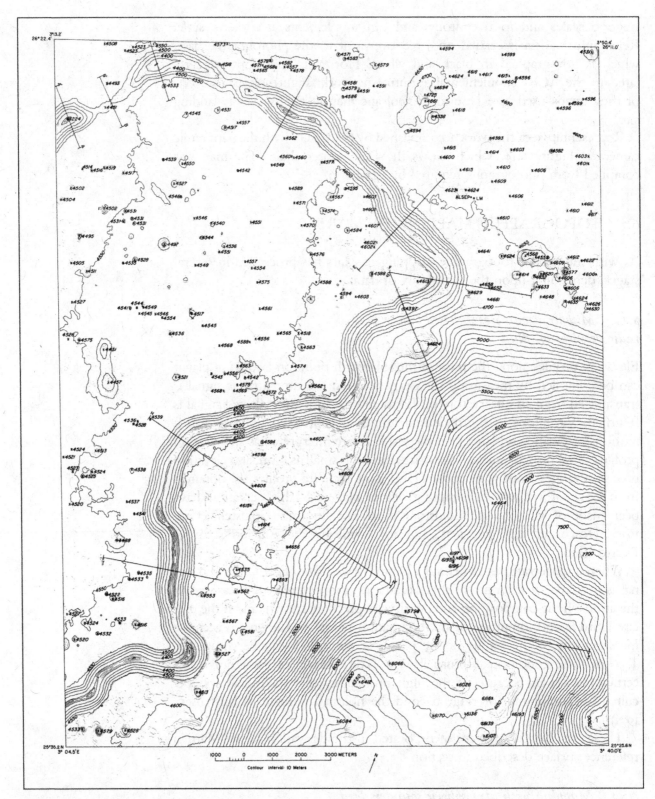

Figure 6.13
Topographic map of Apollo 15 landing site at Hadley Rille compiled from panoramic camera photographs and used as a base for geologic mapping.

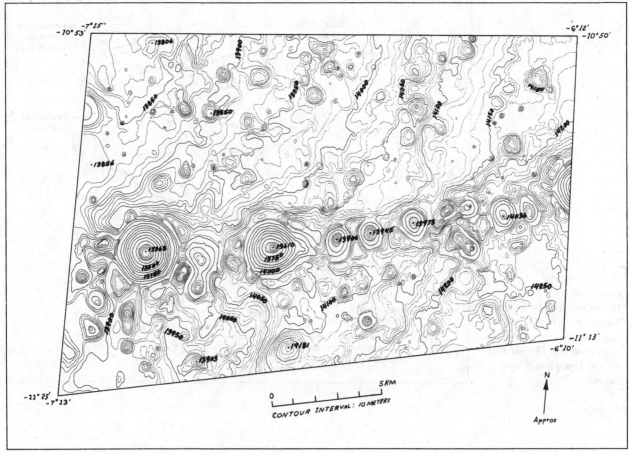

Figure 6.14
Topographic map of Davy crater
chain compiled at 1:50,000 scale
with 10-m contours from panoramic camera photographs acquired by Apollo 16.

spheric experiments. The UVS data were used to determine atmospheric pressure, as well as the zero reference for elevations, described in Section 6.3.4.

The IRIS provided similar information. Herr et al. (1970) and Hanel et al. (1970, 1972) computed elevations with stated reliabilities of 0.5 km for local variations and with absolute accuracies of ±1 km. The Infrared Radiometer (IRR) was used to infer thermal properties of the Martian surface (Chase et al., 1972); these data were also used to compile a temperature map that can be correlated with topographic variations (Cunningham and Schurmeier, 1969). The infrared experiment of Mariner 9 provided about forty-six hundred elevation points, covering virtually the same region as that covered by the UVS. Agreement between the various datasets is within 2 to 3 km.

All of the measurements varied in complicated ways, depending on weather conditions on Mars. Although the profiles are reasonably accurate locally, they required adjustment to occultation and radar measurements, as discussed in Section 6.7.1. Topographic maps of Mars based on these data were compiled by interpolation at scales of 1:25,000,000 and 1:5,000,000 (Wu, 1975, 1978).

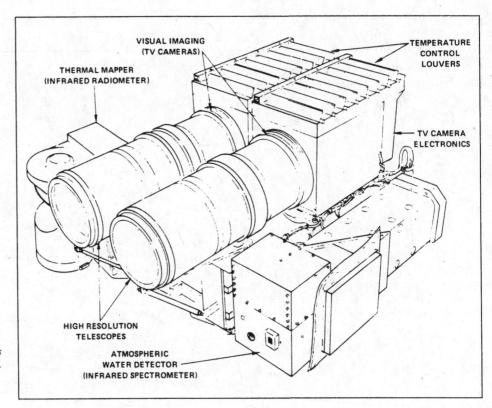

VISUAL IMAGING
(TV CAMERAS)

TEMPERATURE
CONTROL
LOUVERS

THERMAL MAPPER
(INFRARED RADIOMETER)

TV CAMERA
ELECTRONICS

HIGH RESOLUTION
TELESCOPES

ATMOSPHERIC
WATER DETECTOR
(INFRARED SPECTROMETER)

Figure 6.15
The orbiter science platform on Viking 1 and 2 Orbiters. It contained the two television cameras together with the Thermal Mapper and Atmospheric Water Detector.

6.7.3. *Photogrammetric image mapping*

In 1964 Mariner 4 returned twenty-one low-resolution television pictures from Mars. Mariners 6 and 7 also were flyby missions, and together they acquired 201 frames, with a maximum resolution of 500 m. Mariner 9, in 1971, was the first spacecraft to go into orbit around another planet. It acquired seventy-three hundred pictures that cover the entire surface at resolutions of 1 to 3 km. Although these pictures show the extreme diversity of the Martian surface, it was not until 1976 that the Viking 1 and 2 Orbiters obtained coverage from which topographic maps could be produced. Each of these Orbiters carried two identical vidicon cameras (Figure 6.15). The cameras, along with instruments to measure the surface temperature and amount of water vapor in the atmosphere, were mounted on the science platform, a device that could be moved about two axes to achieve a scanning motion when viewing the Martian surface from orbit.

Each camera had a 475-mm focal length telephoto lens with a field-of-view of 1.54 × 1.69 degrees. From an orbital altitude of 1,500 km, each frame covered a minimum surface area of 40 × 44 km. Six selectable filters allow the acquisition of color images. Acquisition of a frame and subsequent readout to the tape recorder required 8.96 seconds, so a frame was taken on alternate cameras every 4.48 seconds. This alternating pattern, coupled with motion along the orbit, produced a swath of two adjacent images.

The image on the vidicon was scanned, or read out, as 1,056 horizontal lines. Each line, in turn, was divided into 1,182 pixels (picture elements), and the brightness of each pixel ranged from 0 to 127 arbitrary units. Thus, to record a single frame required the storage of almost 10 million bits (binary digits) on the orbiter's tape recorder. The pictures were stored on the tape recorder in digital form until there was an opportunity to play back the data over the orbiter's communications system to a receiving station on Earth. The two Orbiters returned over forty-six thousand images in all seasons, lighting, and weather. The coverage, however, is not in uniform systematic patterns as is usually desired for mapping. The spacecraft were in elliptical orbits, and pictures were taken from altitudes ranging from 300 to 33,000 km. Coverage patterns are shown in Tyner and Carroll (1983).

From this diversified set of images, with the secondary control net described in Section 6.4.3, Mars' topography is being systematically mapped at 1:2,000,000 scale with 1-km contours (Wu, 1979). An example of the more than eighty quadrangles produced so far is shown in Figure 6.16. In addition, compilation of a 1:15,000,000-scale global topographic map has been published.

6.7.4. *Viking lander mapping*

The two Viking lander spacecraft each carried two facsimile camera systems located 0.82 m apart and 1.3 m above the nominal landing plane. The mirror scanning device could acquire data from 40 degrees above the horizon to 60 degrees below. The cameras could be rotated 360 degrees about their vertical axis so that they could photograph the scene out to the horizon. The sensors were twelve light-sensitive photo diodes with two sizes corresponding to 0.12- and 0.04-degree instantaneous field of view. In the course of the mission, practically all of the visible panorama was imaged in high resolution at different times of day by both cameras of both landers. Thereby virtually all of the scene was imaged stereoscopically.

A stereoscopic mapping system was developed jointly by Stanford University and the Jet Propulsion Laboratory (Liebes and Schwartz, 1977). In principle, this instrument was equivalent to the analytical stereoplotters described in Section 6.5.1. However, rather than producing hard-copy images, it was designed to use two video monitors that could be viewed through a scanning stereoscope. The contouring function was equivalent to that performed by the analytical plotter operator. The resulting map products are contour plots that can be superposed on the images. Because of the low point of view of the cameras, many of the contours are occluded by rocks and elevation differences within the scenes. Maps were produced at scales between 1:1 and 1:2,000.

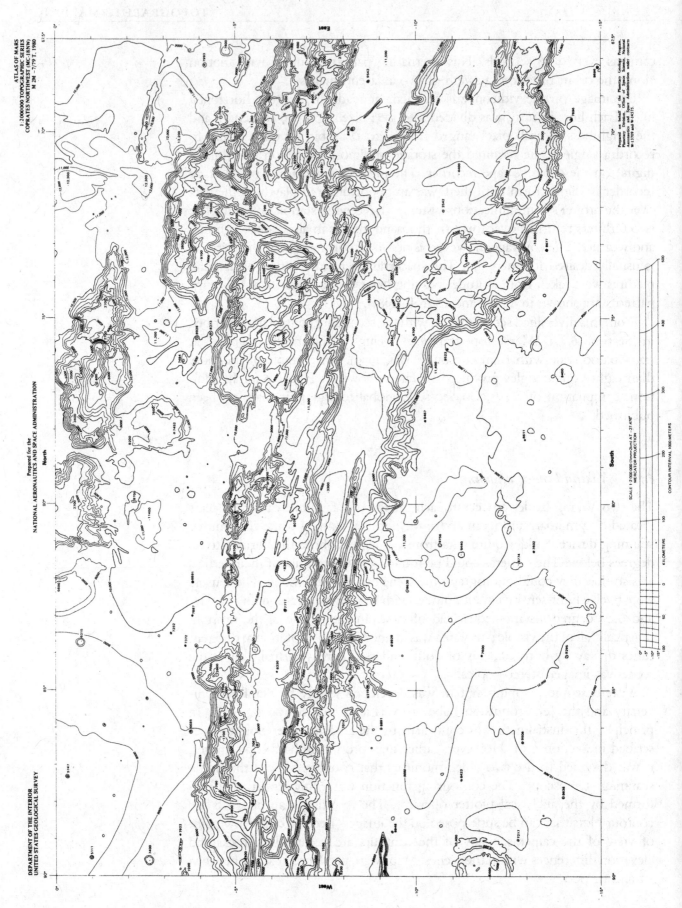

Prepared for the
NATIONAL AERONAUTICS AND SPACE ADMINISTRATION

Prepared on behalf of the Planetary Geology Program,
Office of Space Science, National Aeronautics and Space
Administration under contracts
W-13709 and W-14575

DEPARTMENT OF THE INTERIOR
UNITED STATES GEOLOGICAL SURVEY

North

South

SCALE 1:2 000 000 (1mm=2km) AT -27.476°
MERCATOR PROJECTION
CONTOUR INTERVAL 1000 METERS

KILOMETERS

An explanation of the system and the operating procedures and examples of the maps are given in Liebes (1982).

6.8.   TOPOGRAPHIC MAPPING OF VENUS

The surface of Venus is perpetually shrouded in clouds, and hence it cannot be recorded by sensors operating in the visible wavelengths. Earth-based radar measurements have been made from Goldstone radar in California (Campbell et al., 1972) to produce topographic profiles, using the same technique employed for Mars. The most significant topographic data, however, have been obtained from orbiting radar altimeters carried on Pioneer Venus (Pettengill et al., 1980) and on the USSR's Venera 15 and 16 (Barsukov et al., 1984). The accuracy of spot elevations computed from altimeter observations depends upon the size of the radar footprint and the roughness of the surface; these elevations do not necessarily represent the average elevation value within the footprint. The locations of radar profiles are derived from spacecraft tracking data. The radar altimeter data were useful for defining the general shape of Venus and for delineating continental land masses, but not surface landforms. A global topographic map has been compiled from the Pioneer Venus radar altimetry, and a color-coded relief representation is shown in Figure 6.17.

6.9.   TOPOGRAPHIC MAPPING OF THE SATELLITES OF URANUS

In January 1986 Voyager 2 took a series of pictures of Uranus and its five major satellites. The Voyager Imaging Science System (ISS) (Smith et al., 1986) had two cameras: one wide-angle, of 200-mm focal length, and the other narrow-angle, of 1,500-mm focal length. Each frame of the cameras consisted of 800 × 800 elements with a pixel size of 14 μm. Both cameras were precisely calibrated (Benesh and Jepsen, 1978). A topographic map of the southern hemisphere of Miranda, the smallest and innermost of Uranus' major satellites, was compiled from six stereoimages from the narrow-angle camera (Wu et al., 1987).

6.10.   DIGITAL TOPOGRAPHIC MAPS

The traditional form of publication for topographic information is the contour map, and it is unlikely that this format will become obsolete. Digital maps, however, are becoming indispensable to planetary researchers, who require data that are compatible with their own digital compilations of geophysical, geochemical, and geological information (Batson, 1987).

A digital topographic map can be compiled simultaneously with the com-

Figure 6.16
Facing page: *Topographic map of the Coprates Northwest quadrangle of Mars compiled at 1:2,000,000 scale and 1-km contours from Viking 1 and 2 images.*

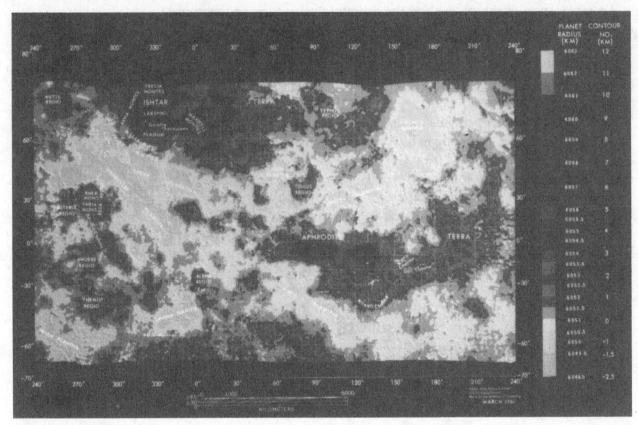

Figure 6.17
Global topographic map of Venus compiled from radar altimetry, with elevations represented by variations in the planetary radius.

pilation of contour maps by preserving the digital files generated by analytical stereoplotters. Such files consist of strings of map coordinates labeled by elevation value; they must be converted to a raster digital terrain model (DTM) to be useful as base maps. A digital terrain model (DTM) resembles a digital image model (DIM), as described in Chapter 3, except that it is an array of elevations rather than of brightness values. A digitized contour map is converted to a DTM by interpolating elevation values between contour lines. A variety of algorithms have been developed for this purpose by numerous organizations that are involved in the compilation of DTMs of terrestrial topographic data. The system used for planetary mapping was developed by Kathleen Edwards, USGS (written commun., 1981), and is very similar to that described by Leberl and Olson (1982). The principle of the algorithm is to interpolate linearly between adjacent contour lines so that all pixels in the array have an assigned value. DTMs are prepared in a sinusoidal map projection for compatibility with the DIMs described in Chapter 3.

Once topographic data are available in digital form, they can be combined with other data types, including images, to produce a variety of useful products, e.g., the perspective view of part of Miranda shown in Figure 6.18.

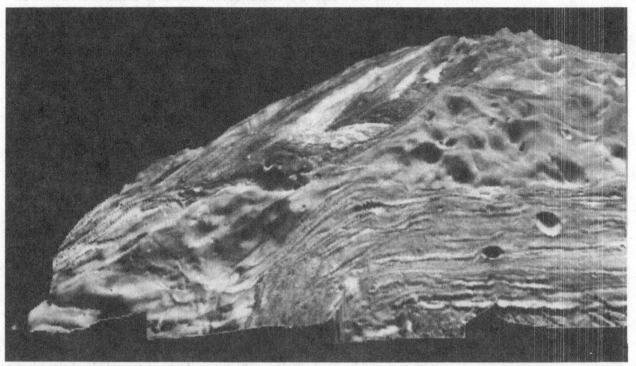

Figure 6.18
Perspective view of part of the southern hemisphere of Miranda, prepared by combining digital terrain data with digital image data and subjecting both to a perspective projection. Width of view is about 300 km.

6.11. TOPOGRAPHIC MAPPING WITH DATA FROM FUTURE PLANETARY MISSIONS

Future planetary orbital missions will carry both visible and radar imaging systems and radar altimeters. The three missions approved to date are Magellan to Venus, Mars Observer, and Galileo to the Jovian system.

6.11.1. *Radargrammetry for Venus*

The Magellan spacecraft to Venus in 1990 will carry a synthetic-aperture, side-looking radar (SAR) and a radar altimeter with a vertical resolution of approximately 15 m. Eighty percent of the surface of Venus will be imaged by SAR during the nominal 243-day mission. The images will provide data for planimetric mapping, and the altimetry will provide topographic measurements. It is likely that the mission will be considerably extended, which may allow image mapping of the remaining 20 percent of Venus' surface, as well as gravity mapping and the gathering of stereoradar images that can be used to make topographic maps. However, spacecraft tracking cannot be done with sufficient precision, at least during the nominal mission, to allow appreciable improvement of the currently adopted spherical figure.

The geometry of radar images is completely different from that of conventional photographic imaging systems. Radar measures distance from the antenna to the imaged points, whereas photographs measure the angles subtended at the camera by the imaged points. Mapping accuracy is affected by factors including image resolution, terrain type, sensor position and orbital parameters, and the geometry of the stereomodel. Compilation software is being developed for the AS–11 AM analytical stereoplotter to accommodate the radar geometry. This software can use either slant-range or ground-range image pairs taken from either the same side or opposite sides of the area to be mapped (Blackwall, 1981). The ongoing development of photogrammetric techniques to mapping with radar will improve mapping accuracy by incorporating higher-order mathematical terms in the stereoplotter software. These techniques are expected to be fully operational in 1990, in time to use stereoscopic SAR images returned by Magellan to make topographic maps of Venus. By then we should also be able to make contour maps using digital steroradar images rather than hard-copy printouts.

6.11.2. *Mars Observer mapping*

Mars Observer is scheduled for launch in 1992; one of its primary goals is to define the global topography and gravitational field of Mars. Topographic measurements will be made by laser altimetry. The entire surface of the planet will be mapped at this resolution during a single Martian year (approximately 2 Earth years). A high-resolution imaging system operating simultaneously with the laser altimeter will be used to locate the laser profiles on existing DIMs that were prepared from Viking Orbiter data.

The Mars Observer altimetry will be formatted as a DTM, and it will provide a dataset that complements the photogrammetric compilations from Viking Orbiter. The gravitational field, and hence the equipotential surface of Mars, will be measured by tracking the spacecraft's orbit and will be incorporated in the DTM. The resulting regional accuracy of the Mars Observer data is therefore expected to be much higher than that of the existing topographic map. The existing map, on the other hand, discriminates topographic details at higher resolution than can be achieved with altimetry. Together, the two datasets should produce the most precise and comprehensive map ever made of an entire planet other than Earth.

6.11.3. *Galileo stereomapping*

Galileo was launched toward Jupiter in late 1989. Although the spacecraft carries an imaging system, mapping is not a primary goal. Images will be acquired of the areas on Jupiter where detached probes will penetrate the atmosphere and of the Jovian satellites during close flybys. The Galileo cameras will not use television vidicons, but rather charge coupled devices (CCDs). These are arrays of light-sensing diodes that are mechanically emplaced in the detector array. They produce images that have higher geometric integrity than vidicons (Klaasen et al., 1982). However, the photogrammetric problems resulting from small formats and narrow-angle lenses remain. The photogrammetric reduction of the images will not differ substantially from that employed with Viking Orbiter data.

6.12. REFERENCES

American Society of Photogrammetry. 1980. *Manual of Photogrammetry,* 4th ed., Chap. X, XI, XII, 519–697.

Arthur, D. W. G. 1962. Model formation with narrow-angle photography. *Photogrammetric Record* 4 (19): 49–53.

Arthur, D. W. G. 1968. A new secondary selenodetic triangulation. *Communication, Lunar and Planetary Laboratory, Univ. of Arizona* 7–6: 313–60.

Arthur, D. W. G., Batson, R. M., and Doyle, F. J. 1970. Geodesy and cartography: A strategy for the geologic exploration of the planets. *U.S. Geological Survey Circular* 640: 12–20.

Barth, C. A., and Hord, C. W. 1971. Mariner ultraviolet spectrometer: Topography and polar cap. *Science* 173: 197–201.

Barsukov, V. L., Basilevsky, A. T., Kuzmin, R. O., et al. 1984. Geologiya Venery rezultatum analiza radiolakatsionnykh izobrazhenii, poluchennykh AMC Venera–16 i Venera–16 (predvaritelnye dannye). *Geokhimiya* 12: 1811–20.

Batson, R. M. 1987. Digital cartography of the planets: New methods, its status, and future. *Photogrammetric Engineering and Remote Sensing* 53 (9): 1211–18.

Benesh, M., and Jepsen, P. 1978. Voyager Imaging Science System calibration report. *Jet Propulsion Laboratory* 618–802.

Bill, G. G., and Ferrari, A. J. 1977. A harmonic analysis of lunar photography. *Icarus* 31: 244–59.

Blackwall, B. H. 1981. Realtime math model from SAR imagery. *Final Technical Report, RADC-TR–301.*

Campbell, D. B., Dyce, L. B., Ingall, R. P., et al. 1972. Venus: Topography revealed by radar data. *Science* 175: 514–16.

Chase, S. C., Jr., Hatzenbeler, H., Kieffer, H. H., et al. 1972. Infrared radiometry experiment on Mariner 9: *Science* 175: 308–9.

Christensen, E. J. 1975. Martian topography derived from occultation radar, spectral and optical measurements. *J. Geophys. Res.* 80: 2909–13.

Cunningham, N. W., and Schurmeier, H. M. 1969. Introduction. *Mariner Mars 1969: A Preliminary Report: NASA SP–225:* 1–36.

Davies, M. E. 1973. Mariner 9: Primary control net. *Photogrammetric Engineering* 39 (12): 1297–1302.

Davies, M. E., Katayama, F. Y., and Roth, J. A. 1978. Control net of Mars. *The Rand Corporation, R–2309-NASA.*

Defense Mapping Agency. 1987. Department of Defense World Geodetic System, 1984. *Defense Mapping Agency Technical Report DMA TR 8350.2,* 30 September 1987.

Downs, G. S., Goldstein, R. M., Green, R. R., and Morris, G. A. 1971. Mars radar observations: A preliminary report: *Science* 174: 1324–7.

Downs, G. S., Goldstein, R. M., Green, R. R., et al. 1973. Martian topography and surface properties as seen by radar: The 1971 opposition. *Icarus* 18: 8–21.

Downs, G. S., Reichley, P. E., and Green, R. R. 1975. Radar measurements of Mars topography and surface properties: The 1971 and 1973 oppositions. *Icarus* 26: 273–312.

Doyle, F. J., Elassal, A. A., and Lucas, J. R. 1977. Selenocentric geodetic reference system. *NOAA Technical Report NOS 70 NGS 5.*

Elassal, A. A. 1976. General Integrated Analytical Triangulation (GIANT) program. *USGS Report W5346.* U.S. Geological Survey, Department of Interior, Reston, Va.

Elassal, A. A., and Malhotra, R. C. 1987. General integrated analytical triangulation

(GIANT) user's guide. *NOAA Technical Report NOS 126, CGS 161.*

Ferrari, A. J., Sinclair, W. S., Sjogren, W. L., et al. 1980. Geophysical parameters of the Earth–Moon system: *J. Geophys. Res.* 85 (B7): 3939–51.

French, B. M. 1977. "What is new on the Moon-II." *Sky and Telescope* 53 (4): 257–61.

Goldstein, R. M., Melbourne, W. G., Morris, G. A., et al. 1970. Preliminary radar results of Mars. *Radio Science* 5: 475–8.

Hanel, R. A., Conrath, B. J., Hovis, W. A., et al. 1970. Infrared spectroscopy experiment for Mariner, 1971. *Icarus* 12: 48–62.

Hanel, R. A., Conrath, B. J., Hovis, W. A., et al. 1972. Infrared spectroscopy experiment on Mariner 9 mission. *Science* 175: 305–8.

Herr, K. C., Horn, D., McAfee, J. M., and Pimentel, G. C. 1970. Martian topography from the Mariner 6 and 7 infrared spectra. *Astronomical Journal* 75: 883–94.

Jordan, J. F., and Lorrell, J. 1975. Mariner 9: An instrument of dynamical science. *Icarus* 25: 146–65.

Klaasen, K. P., Clary, M. C., and Janesick, J. R., 1982. Charge-coupled device (CCD) television camera for NASA's Galileo mission to Jupiter. *SPIE 331 Instruments in Astronomy IV:* 376–87.

Kliore, A. J., Cain, D. L., Fjeldbo, G., and Seidel, B. L., 1972. Mariner 9 S-band martian occultation experiment. *Science* 175: 313–17.

Kliore, A. J., Fjeldbo, G., Seidel, B. L., et al. 1973. S-band radio occultation measurements of the atmosphere and topography of Mars with Mariner 9: Extended mission coverage of polar and intermediate latitudes: *J. Geophys. Res.* 78: 4331–51.

Kopal, Z., and Carder, R. W., 1974. Mapping the Moon: Past and present. *Astrophysic and Space Science Library,* vol. 50. D. Reidel.

Leberl, F. W., and Olson, D., 1982. Raster scanning for operational digitizing of graphic data. *Photogrammetric Engineering and Remote Sensing* 4: 615–27.

Liebes, S., 1982. Viking Lander atlas of Mars. *NASA Contractor Report 3568.*

Liebes, S., and Schwartz, A. 1977. Viking 1975 Mars Lander interactive computerized video stereophotogrammetry. *J. Geophys. Res.* 82: 4421–9.

Lindal, G. F., Hotz, H. B., Sweetnam, D. M., et al. 1979. Viking radio occultation measurements of the atmosphere and topography of Mars. *J. Geophys. Res.* 84 (B14): 8443–56.

Lorell, J., Anderson, J. D., and Shapiro, I. I. 1970. Celestial mechanics experiment for Mariner Mars 1971. *Icarus* 12: 78–81.

Lorell, J., Born, G. H., Jordan, J. F., et al. 1972. Mariner 9 celestial mechanics experiment: Gravity field and pole direction of Mars. *Science* 175: 317–20.

Meyer, D. I., and Ruffin, B. W. 1965. ACIC selenodetic control system. *Icarus* 4: 513–27.

Moore, H. J., and Lugn, R. 1966. Experimental topographic map of small area of the lunar surface from the Ranger VIII photographs. *Technical Report No. 32–800: Ranger VIII and IX, National Aeronautics ands Space Administration/Jet Propulsion Laboratory:* 295–302.

Pettengill, G. H., Counselman, C. C., Rainville, L. P., and Shapiro, I. I. 1969. Radar measurements of martian topography. *Astron. J.* 74: 461–82.

Pettsengill, G. H., Rogers, A. E. E., and Shapiro, I. I. 1971. Martian craters and a scarp as seen by radar. *Science,* 174: 1321–4.

Pettengill, G. H., Eliason, E., Ford, P. G., et al. 1980. Pioneer Venus radar results: Altimetry and surface properties. *J. Geophys. Res.* 85: 8261–70.

Rogers, A. E. E., Ash, M. E., Counselman, C. C., et al. 1970. Radar measurements of the surface topography and roughness. *Radio Science* 5: 465–673.

Schimerman, L. A., 1973. *NASA Lunar Cartographic Dossier* 1: 3.1–3.5.

Sjogren, W. L., Ananda, M., Williams, B. G., et al. 1981. Venus gravity fields. *Annales de Geophysique, Centre National de la Recherche Scientifique* 37 (1): 179–85.

Slama, C. C., Ebner, H., and Fritz, L. 1980. Aerotriangulation. *Manual of Photogrammetry,* 4th ed., American Society of Photogrammetry, pp. 453–518.

Smith, B. A., Soderblom, L. A., Beebe, R., et al. 1986. Voyager 2 in the Uranian system: Imaging science results. *Science* 233: 43–64.

Tyner, R. L., and Carroll, R. D., 1983. A catalog of selected Viking Orbiter images. *NASA Reference Publication 1093*.

Wu, S. S. C. 1969. Photogrammetry of Apollo 8: Analysis of Apollo 8 photography and visual observations. *Apollo 8 Preliminary Science Report, National Aeronautics and Space Administration SP-201*: 33–4.

Wu, S. S. C. 1975. Topographic mapping of Mars: *U.S. Geological Survey Interagency Report: Astrogeology 63*.

Wu, S. S. C. 1978. Mars synthetic topographic mapping. *Icarus 33*: 417–40.

Wu, S. S. C. 1979. Contour mapping: *Atlas of Mars, 1:5,000,000 map series, National Aeronautics and Space Administration SP-483*, 130–7.

Wu, S. S. C., 1981a. A method of defining topographic datums of planetary bodies. *Annales de Geophysique, Centre National de la Recherche Scientifique 37* (1): 147–60.

Wu, S. S. C. 1981b. New global topographic mapping of the Moon. *Proceedings, 12th Lunar and Planetary Science Conference*: 1217–18.

Wu, S. S. C., 1985. Topographic mapping of the Moon. *Earth, Moon, and Planets 32* (2): 165–72.

Wu, S. S. C., Elassal, A. A., Jordan, R., and Schafer, F. J. 1982. Photogrammetric application of Viking orbital photography: *Planetary and Space Science 30* (1): 45–55.

Wu, S. S. C., Schafer, F. J., Jordan, R., and Howington, A. E., 1987. Topographic map of Miranda. *Lunar and Planetary Science XVIII, LPI/USRA*: 110–11.

7

Geologic mapping

DON E. WILHELMS

> "On first examining a new district [planet] nothing can appear
> more hopeless than the chaos of rocks [landforms]; but by
> recording the stratification and nature of the rocks and fossils
> [morphologic patterns] at many points, always reasoning
> and predicting what will be found elsewhere, light soon begins
> to dawn on the district [planet], and the structure of the whole
> becomes more or less intelligible."
>
> Darwin (1958), p. 77
> (called to my attention and paraphrased by John F. McCauley).

7.1. INTRODUCTION

Many scientists as well as laymen are surprised to learn that geologists study
the Moon and planets. The surprise may abate when they consider that these
extraterrestrial bodies consist of rock. But both the layman and the uninitiated
professional geologist may find it harder to understand how planetary crusts
can be deciphered without direct field study or sampling. In fact, the basic
architecture and postformational history of the impact-scarred, relatively stable
Moon, Mercury, Callisto, and Martian and Saturnian satellites, the tectonically
deformed Ganymede, the constantly changing Io, and the richly diverse Mars
have been determined on the basis of remotely obtained data.*

A major tool for unravelling the secrets of these distant objects has been
geologic mapping. A geologic map is a two-dimensional representation of
the three-dimensional spatial relations and chronologic sequences of the materials and structures of a planetary crust. In constructing a map, the geologist
organizes and summarizes a seemingly bewildering planetary scene into a
comprehensible picture. A properly constructed map distinguishes the planet's
essential rock framework from irrelevant detail. Like the graphs of more quan-

---

* Henceforth in this chapter, the term "planet" includes satellites as well as true planets.

I am much indebted to B. K. Lucchitta for her thorough critical review of an earlier draft
of this chapter. R. J. Baldwin, R. M. Batson, J. E. Guest, and C. J. Hayden also contributed
valuable comments.

titative sciences, geologic maps are economical media for documenting the observations of the mapper and communicating the results to others.

The success of planetary geologic mapping rests on the analytical power of the uniquely geologic perspective on which it is based: the perspective of *history* (Albritton, 1963). Fundamental to geology is the science of *stratigraphy,* which determines the sequence of emplacement of the planet's three-dimensional building blocks, its *geologic units.* Geologic maps also show the age relations of these material units to the structures created by *tectonic* deformation by internal forces. The principles that guide this mental look beneath the surface and back in time are elementary: that younger rock units usually lie on older rock units and structures and that younger structures and erosional surfaces cut across older rock units, structures, and erosional surfaces. These principles can be applied not only to Earth but to the solid surfaces of all planets because the relevant geometric relations can commonly be observed on images. Geologists accustomed to physical contact with the rocks of Earth can readily transfer their basic approach to other planets once they have learned the peculiarities of the new data set. There are no separate sciences like "selenology" or "ganymedology."

This chapter outlines the general principles and methods by which geologic maps of the planets are constructed from images. It also briefly shows how other remote sensing data are incorporated into the mapping. Section 7.2 summarizes the principles of geologic mapping for a general audience. Some familiarity with geologic principles and with the planets is helpful but not essential to the reading of this section. Section 7.3 briefly discusses some difficulties that have been encountered in viewing planetary images. Section 7.4 presents detailed guidelines for the construction and review of maps for those who intend to map actively.

Most examples are taken from lunar mapping because extraterrestrial mapping techniques were evolved for this second geologically mapped planet and because the lunar techniques, in turn, have been adapted to the other planets. The lunar and general examples are based largely on several earlier papers by the author and his colleagues in the U.S. Geological Survey's Branch of Astrogeology (Shoemaker, 1962a; Shoemaker and Hackman, 1962; McCauley, 1967; Wilhelms, 1970, 1972, 1980, 1987; Wilhelms and McCauley, 1971; Carr, 1984). These studies were almost entirely funded by the U.S. National Aeronautics and Space Administration (NASA). Section 7.4, in particular, draws heavily on an informal document prepared in 1972 for the use of geologic mappers participating in the program of lunar and planetary mapping conducted for NASA under the guidance of the Geological Survey (Wilhelms, 1972). A perceptive analysis of the logic of geologic maps in general is given by Varnes (1974). Guest and Greeley (1977) have contributed many insights about geologic mapping in their summary of lunar geology. This chapter and all other considerations of extraterrestrial stratigraphy and geologic mapping also owe a great debt to the lucid presentation by the late and much mourned Tim Mutch (Mutch, 1970).

Statements of scientific fact or interpretation given in this chapter are intended only to illustrate the mapping and are not fully supported or referenced here. The reader interested in their background and further reading is referred to books edited by Morrison (1982), for the satellites of Jupiter, and by Carr (1984), for the terrestrial planets.

7.2. RATIONALE AND GENERAL METHODS

7.2.1. *The geologic unit*

The process of geologic mapping is simple in principle. It rests on the fundamental concept that a planetary crust is composed of discrete three-dimensional bodies of rock called geologic units. The rocks of each unit formed, relative to those of the neighboring units, (a) by a discrete process or related processes and (b) in a discrete timespan. The unit concept reduces the complex internal detail of each body of rock to a more comprehensible entity.

On Earth geologic units are detected in vertical exposures or on the ground surface, where they can be reconstructed from outcrops or soil fragments. Vertical sections are also sometimes observed on Mars but very rarely on the other planets. Certain lunar geologic units have been mapped on the surface by means of the overlying fragmental material *(regolith)* that was derived from the bedrock units. Mostly, however, planetary geologic units are detected by their topographic expression and other remotely observed surface properties. Despite the differences in data and geologic style among planets, the procedures for recognizing geologic units are basically the same for all planets.

Planetary mapping can begin when a mapper decides that a certain terrain has been formed by a uniform set of processes and in a specific timespan, even if the exact processes and absolute time are unknown. The mapper looks for distinctive surface morphologies (coarse features) or textures (fine features) that are similar or regularly gradational over the whole terrain. For example, a flow lobe, or a repetitive pattern of flow lobes of the same general size and shape, probably reflects a unit (Figure 7.1). The lobes may consist of lava, but this interpretation is not essential to identification of the unit. A common type of gradational deposits are those that surround the innumerable craters observed on planets. The deposits of many craters consist of rough, concentrically textured material near the crater rim, grading through radial ridges or lobes farther out, to chains, loops, or clusters of pits at greater distances (Figure 7.2). The radial and concentric arrangement of these textures shows that they are related to the crater. Studies have shown, further, that the crater deposits contain *ejecta* that was excavated from the crater during the *primary* impact of a projectile from space. The small satellitic pits are *secondary-impact* craters formed by impacts of this primary ejecta. Deposits of the secondaries are also geologic units that may be mixed with the primary ejecta. Again,

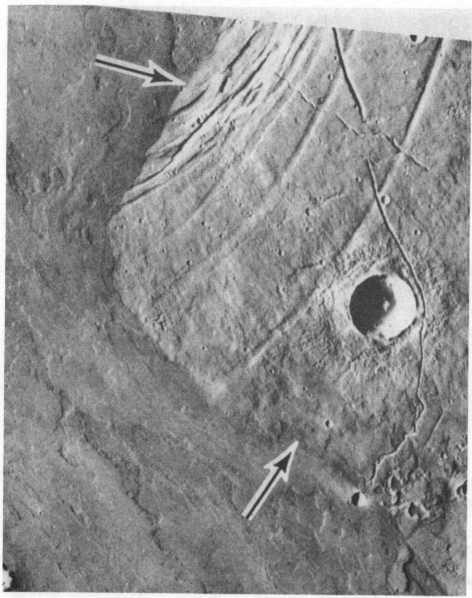

Figure 7.1
Distinct contacts between geologic units on Mars. Upper arrow, scarp contact; lower arrow, pinchout contact.
Lobes of younger unit (left) suggest lava origin. The left sector of the crater's ejecta subdues an older, broad
fault graben, whereas the right sector is cut by a younger, sharp fault graben. The relative ages demonstrated
by these relations are confirmed above the crater, where the sharp graben cuts the broad graben. Viking orbiter
frame 623A09, 99 km high by 82 km wide, centered at 26.6° N, 127.4° W.

these interpretations are not essential to the recognition of a crater deposit as
a unit, although they help considerably in refined mapping, as will be shown.
No matter what its origin, the whole textured area around the crater is the
expression of a complex geologic unit that consists of horizontally gradational
facies (physically different parts of the same deposit).

A

Figure 7.2
Stratigraphic relations near lunar crater Euler (E; 28-km diameter; 23.3° N, 29.2° W) in southern Mare
Imbrium. A. Photograph. Lobate, channeled mare lavas evident to left (west) of Euler. These and other young
mare materials truncate the deposits and secondary craters of Euler (arrows), whereas the Euler materials
northeast of the crater are completely developed and superposed on an older mare unit. A belt of mare at lower
arrow, containing sinuous rille, separates an outlying fan of Euler ejecta (e) from the main outcrop. Mare
surface sharply abuts smaller crater (white arrowhead, lower left) and mountains (parts of Imbrium basin). C,
secondary craters of Copernicus, centered 450 km to SSE, superposed on Euler and mare deposits; the most
conspicuous craters of each secondary cluster lie closest to Copernicus. Apollo 17 mapping frame 2293.
(Continued)

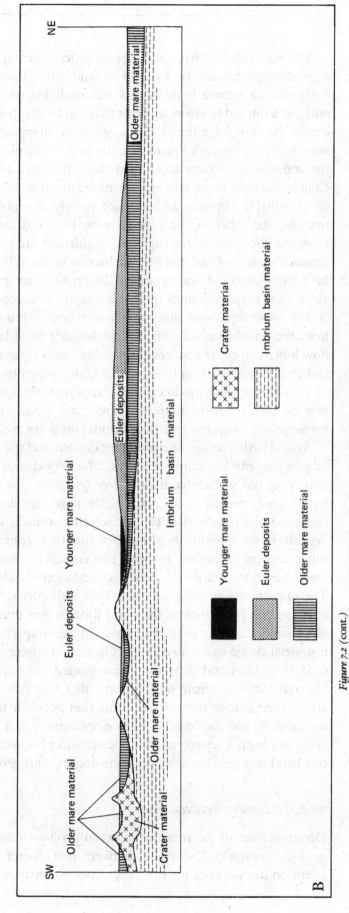

Figure 7.2 (cont.)

B. Diagrammatic geologic cross section drawn from lower left (southwest) corner to upper right (northeast) corner of 7.2A. Basement unit is Imbrium-basin material, which crops out in the mountains. The crater in the southwest is superposed on this basement. This crater and the rest of the Imbrium-basin basement are flooded by the older mare material. Euler deposits are superposed on this mare material and drape over parts of the Imbrium mountains. Parts of the Euler deposits are flooded by the narrow belt of younger mare material (lower arrow in A).

213

Although detected from surface properties, a geologic unit is not a surface, a geomorphic terrain, or a group of landforms. Instead, it is the sheetlike, wedgelike, or tabular body of rock that underlies the surface. Each geologic unit has a limited horizontal extent that can be mapped and a limited vertical extent that can be inferred. Thus, geologic mapping concerns the layered section of rock beneath a planar area more than the plain itself, and it concerns the deposits of a crater more than the cup-shaped topographic depression. Craters and their larger relatives, the multiringed impact basins, are the sources of identifiable deposits and are not merely assemblages of rings, ridges, troughs, and other topographic features. Even old, largely undistinctive surfaces like the lunar terrae (uplands, highlands) are not underlain by homogeneous masses of material but by discrete units, although such terrains may have to be assigned noncommittal designations temporarily, such as "undivided terra material," until their stratigraphy is learned (see Section 7.2.7).

The most significant units are those whose defining characteristics result from the units' emplacement process and not from later modifications. The flow lobes and ejecta textures already illustrated are examples of such intrinsic textures. Erosional morphology, faults, and superposed craters are examples of postdepositional modifications. Properties resulting from modifications may be used to define units (Sections 7.2.3, 7.2.5, 7.2.6), but are subject to reassessment when the true depositional units are recognized.

As on Earth, the age of a planetary deposit and the ages of its components may or may not be entirely the same. Most flood lavas or pyroclastic blankets that cover old surfaces constitute geologic units that consist of juvenile materials newly melted in the mantle. Absolute ages determined from samples of these units usually date the surface emplacement of the unit. An impact deposit, however, consists of material that was created before the impact by other impacts, volcanism, or other processes. The projectile that formed each crater penetrated, violently disrupted, and ejected older units from its target. The impact may or may not "reset" the radiometric "clocks" of these older components. Nevertheless, the ejecta forms a new unit when it falls upon the older surface surrounding the crater. In this respect, impact ejecta is like a terrestrial clastic sedimentary rock. On planets where water and ice are agents of both erosion and deposition, transported materials similarly contain old material but may create new deposits that take new stratigraphic positions above their source units. This distinction between the "origin" of a unit's constituents and the "origin" of its emplacement as a three-dimensional rock body has been a source of misunderstanding between stratigraphers on the one hand and geochronologists, petrologists, and geochemists on the other.

7.2.2. Relations between units

Determination of the relations between geologic units is at the heart of the geologic approach. The *contacts* between units therefore command particular attention during geologic mapping. They summarize and abstract many observed details; all the terrain enclosed by a contact has been determined to belong to one unit and the terrain on the other side to a different unit or units. The geometric relations of the contacts reflect the depositional patterns and age relations of the units. They may also reveal much about the nature of the materials of the units they bound. For example, cohesive materials, such as lava flows, are commonly bounded by distinct scarps (Figure 7.1, upper arrow), whereas particulate material grades imperceptibly with the subjacent units. When all contacts are drawn and interpreted, a geologic map is nearly complete.

The presence of two units in contact is generally suggested by topographic contrasts that are not explicable by variations within a single unit. Some of these contrasts are lateral and the contacts between the two adjacent units are abrupt. Termination of a plains deposit against a mountain range or against a crater rim or wall is a common type of abrupt contact (Figure 7.2A, lower left). Contrary to appearances, this abrupt termination does not reflect the abutment of two blocks with limitless depth, but rather the superposition of a layer of younger plains material on other, older layers. As shown in Figure 7.2B, the buried layers consist of crater ejecta and basin material that continue laterally beneath the plains. Subtle flooding of the ridged ejecta or satellitic craters of a crater reveals a similar overlap of the plains material on the crater (Figure 7.2A, black-and-white arrows).

The buried layers are commonly invisible at the surface. Their existence is then inferred either from nearby outcrops or from independent knowledge of the stratigraphy. For instance, an extensive ejecta blanket that has been flooded by younger lava may appear in several unflooded windows. Its identity as a unit is established by the exposure in each window of similar textures, similarly oriented linear features, progressive gradations in texture, etc. (Figure 7.2A, radial ejecta of Euler at e, below the belt of mare).

Other lateral contrasts in topography are gradational, and the underlying unit is partly visible. The textures of the underlying unit may grade from distinct near the contact, where the superposed deposit is thin or absent, to invisible, where the superposed deposit is thick. The deposits of superposed craters are very common gradational units. Their radially ridged ejecta and the underlying topography may both be expressed (Figure 7.3A, letter E). More commonly, the superposed deposit retains no texture of its own and merely smooths the underlying unit. Crater deposits commonly pinch out to a vanishing point that is hard to locate and map precisely. Nevertheless, their superposition on other units is demonstrated where their ejecta is fully developed, with no facies missing (Figure 7.2A, northeast of E).

Similar criteria establish the relations between geologic units and structures. Sharp truncation or total obscuration of a fault by a deposit indicates, of course, that the fault is older than the deposit (Figure 7.1, upper arrow). An older fault may be partly expressed topographically as a sag in an overlying younger deposit (Figure 7.1, left of the crater). Where a deposit terminates abruptly against a fault and does not reappear on the other side, the fault is

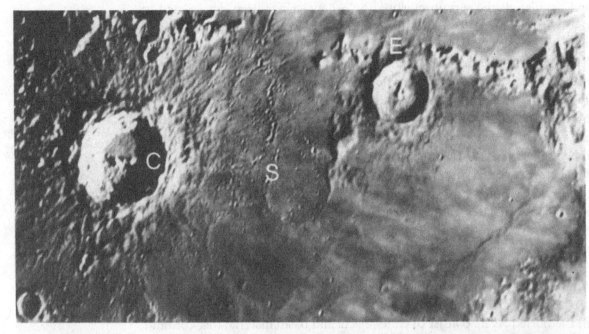

A

B

Figure 7.3
Telescopic photographs of Copernicus–Eratosthenes region of the Moon. See Figure 7.5 for location. A. Low-sun
illumination. C, Copernicus (93-km diameter); E, Eratosthenes; S, crater Stadius and Stadius chain of craters,
extending to top of picture; Montes Apenninus at right (east). At E, deposits of Eratosthenes soften peaks of the
Apennines and extend onto mare surface. Mt. Palomar Observatory photograph. B. High-sun illumination
(near full Moon). Copernicus at left; arrow, Eratosthenes (barely visible). U.S. Naval Observatory photograph.

also probably older than the deposit and has blocked its lateral expansion. Where a sharp fault interrupts a deposit's continuity but does not appear to have influenced the deposit's emplacement, the fault is probably younger than the deposit (Figure 7.1, right of the crater). Intersections of faults may reveal the faults' relative ages (Figure 7.1, above the crater).

These relations illustrate the laws of stratigraphy that were enunciated for Earth by Nicolaus Steno in 1669 (Gilluly et al., 1951): Each new sedimentary deposit (1) was deposited on older deposits and remains above them unless subsequently disturbed (law of superposition), (2) was deposited approximately horizontally and approximately parallel to the underlying surface (law of original horizontality), and (3) spread out laterally until it pinched out or was blocked (law of original continuity). These simple principles, which seem not to have been obvious before 1669 even for Earth, are still worth remembering during planetary work. A test of whether a given terrain consists of one unit is to consider whether the observed morphology can be explained by one laterally continuous rock body. If it cannot, what additional adjacent or subjacent rock bodies, postdepositional deformations, or erosional truncations can explain its appearance? The third dimension should be continuously kept in mind by drawing vertical cross sections.

One variation on these laws encountered in planetary work is that some units, namely secondary-impact craters and the bright rays that emanate from them on many planets, were discontinuous when deposited. Each patch of ray or secondary ejecta obeys Steno's laws, but the patches are collectively treated as a unit. The primary crater that was the source of the secondaries can often be established by mapping or simple inspection of the rays' radial patterns. Units on which the rays or secondaries are superposed are older than the primary crater. Units superposed on the rays or secondaries are younger than the primary. Clusters of secondary craters whose rays are no longer visible can be used similarly because they contain clues to the direction in which the primary lies: The largest secondary craters are concentrated on the side of the cluster nearest the primary, and secondary ejecta with distinctive herringbone textures is concentrated at the opposite side (Figure 7.2A, letter C). Such relations greatly extend the range over which primary impact craters can be dated. The criteria are particularly useful where the primary source is a giant ringed impact basin.

The process of distinguishing significant units and establishing relative ages is well illustrated by the first modern lunar geologic maps that were constructed by stratigraphic principles (see Chapter 2). Mapping of the Copernicus region by Shoemaker (1962a, b; Shoemaker and Hackman, 1962) yielded a clear stratigraphic succession, from oldest to youngest: Imbrium-basin material, mare material, deposits of the crater Eratosthenes, deposits of the crater Copernicus. This sequence can be observed even on a relatively crude photograph that shows the rays of Copernicus crossing Eratosthenes, the deposits and rayless secondary craters of Eratosthenes crossing the adjacent mare surface, and mare embayments abutting Montes Apenninus, which mountains are part

Figure 7.4

Montes Apenninus (rugged range dominating picture), Montes Haemus (MH, lower right), Mare Imbrium (MI), Mare Serenitatis (MS), Mare Vaporum (MV), Palus Putredinis (PP) including Rima Hadley and Apollo 15 landing site (black-and-white arrow), Sinus Aestuum (SA), and craters Archimedes (white arrows), Aristillus (Ar), Autolycus (Au), and Eratosthenes (E, barely visible at lower left; compare Figure 7.3A). White arrows show flooding of Archimedes interior and secondary craters by mare materials; at lower white arrow, Archimedes secondary craters are superposed on nonmare plains deposit, which is also flooded by mare material but which embays Montes Apenninus. See Figure 7.5 for location. Catalina Observatory photograph 1894.

of the rim of the Imbrium basin (Figure 7.3). In a neighboring area, this sequence was augmented by observations of stratigraphic relations near the crater Archimedes (Figure 7.4). Mare material so deeply floods Archimedes in some sectors that only part of the crater wall and rimflank are exposed. Elsewhere, the deposits and secondary craters of Archimedes are well exposed and are superposed on a nonmare plains deposit that, in turn, surrounds or abuts outlying hills of Montes Apenninus (Figure 7.4, lower arrow). The sandwiching of the Archimedes deposits between the mare and nonmare plains materials shows that a finite amount of time elapsed between the formation of the Imbrium basin and the deposition of the mare materials. Thus, the impact that created the basin did not create the mare, which is volcanic basalt (Baldwin, 1949; 1963). This important discovery appeared as novel in an era when the mare and basin were often confused.

Stratigraphic relations in a nearby region convincingly prove this basin– mare distinction and illustrate the power of stratigraphic principles to solve problems of planetary geology (Figure 7.4). Montes Haemus are part of the circular basin that contains Mare Serenitatis. The surface of the mountains is distinctly scoured by grooves and ridges that radiate from Montes Apenninus. Mare Serenitatis is not thus affected. Clearly, therefore, the Imbrium basin postdates Montes Haemus and the rest of the Serenitatis basin but predates Mare Serenitatis (Baldwin, 1949, pp. 210–13). Thus, Mare Serenitatis did not form simultaneously with either the Imbrium basin or the Serenitatis basin. The mare lavas can furthermore be subdivided by overlap relations and crater densities.

A stratigraphic sequence of no fewer than eight major units can be readily constructed in the region of Figures 7.3 and 7.4, from oldest to youngest: (1) Montes Haemus and other Serenitatis basin deposits, (2) the Imbrium basin materials, including those atop Montes Apenninus and those that created the surface lineations of Montes Haemus, (3) the nonmare plains deposit, (4) Archimedes deposits, (5) an older mare unit exposed along the margin of Mare Serenitatis that undoubtedly extends beneath the younger mare of central Serenitatis, (6) the mare units in central Mare Serenitatis and in Palus Putre- dinis between Archimedes and Montes Apenninus, (7) Eratosthenes deposits, and (8) Copernicus deposits (Figure 7.5). These relations can be portrayed in realistic cross sections showing the third dimension (Figure 7.5B). Their validity has been further confirmed by analyses of samples collected by Apol- los 15 (Figure 7.4) and 17 from (1) mountains that are part of the east- ern Serenitatis rim (3.87 ± 0.03 aeons), (2) the foot of Montes Apenninus (3.85 ± 0.03 aeons), (3) part of the border material of Mare Serenitatis (3.72 ± 0.05 aeons), and (4) the mare basalt of Palus Putredinis (3.29 ± 0.05 aeons).*

7.2.3. *Correlations of units*

A major goal of planetary geologic mapping, like terrestrial mapping, is to integrate local stratigraphic sequences ("columns") of geologic units into a stratigraphic column applicable over the whole planet. When any part of this global column is calibrated with absolute ages obtained from samples at spot localities, approximate absolute ages can be matched with the rest of the relative ages (Shoemaker, 1962a; Greeley and Carr, 1976). For example, dating the samples obtained by Apollos 14 and 15 from Imbrium-basin ma-

---

* 1 aeon = 10^9 years.

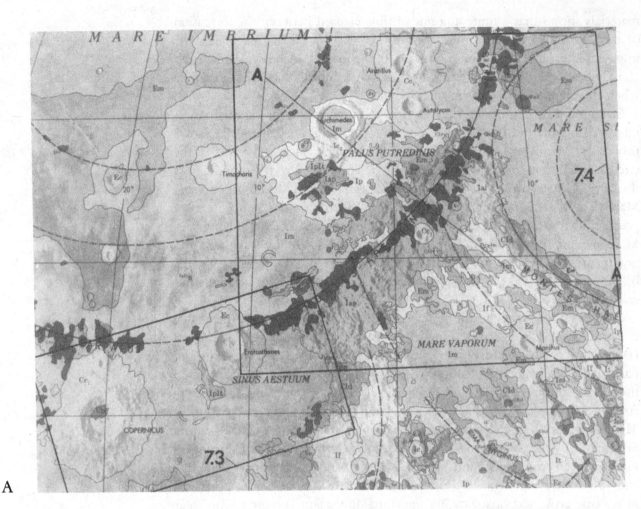

A

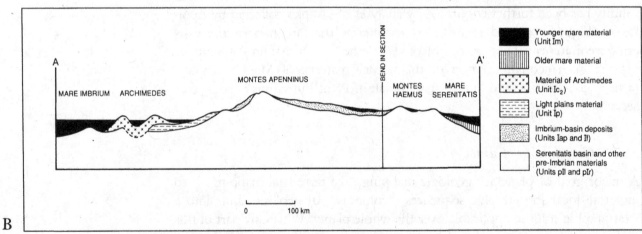

B

Figure 7.5

Geologic relations on part of the lunar nearside. A. Geologic map (Wilhelms and McCauley, 1971) including areas of Figures 7.3 and 7.4. Scale: each 10 degrees of lunar latitude covers 300 km. B. Geologic cross section along line A–A' of 7.5A. The "older mare unit" is shown on the map as Eratosthenian (unit Em) but was demonstrated by Apollo photography to be older than unit Im in Mare Serenitatis. Great vertical exaggeration; lunar curvature ignored.

terial has divided the entire lunar stratigraphic column into two blocks, one older and one younger than the 3.85-aeon age of Imbrium. Samples from other geologically mapped planets will similarly bracket their major historical episodes.

The local stratigraphic columns are generally determined during mapping in the ways that have been discussed. Where possible, local units are dated relatively to extensive units like the deposits and secondary craters of ringed basins, which serve as stratigraphic datum planes for large parts of many planets. However, not all units contact these datum planes. Even some units quite close to basins have been missed by the basin ejecta, which forms lobes of unequal length. Lava flows provide excellent local stratigraphic reference planes but are less useful for regional correlations because of their limited extent. Furthermore, no unit covers an entire planet.

Supplementary means of dating are therefore needed. On Earth, age relations of units that are not in contact can be reconstructed from fossils. On most other planets – though not the ice-covered Europa or the volcanically hyperactive Io – the ages of the units correlate with the density of smaller superposed impact craters (Basaltic Volcanism Study Project, 1981, Chap. 8). Heavily cratered units are usually older than less heavily cratered units (provided that the superposed craters were formed by primary impacts from space, which are spatially almost random and accumulated in proportion to time, and not by secondary impacts, which are concentrated around a primary and formed in a burst when the primary forms). Craters employed as counters in this way are modifiers of other units.

Craters are, of course, also stratigraphic units in their own right. As such, they can be dated by smaller superposed craters. Impact craters on airless planets can also be approximately dated on the basis of their morphology, which is initially similar for a given crater size. Topographically sharp craters are younger than craters that have only bland, rounded topography (Section 7.2.7; Pohn and Offield, 1970; Trask, 1971). Crater morphology is a less satisfactory means of dating than crater frequency, but it has been used successfully on the Moon and Mercury. It is a less reliable guide to age on Mars, where more diverse degradational processes have operated and where crater morphology also depends on latitude and target material (volatile-rich or volatile-poor).

7.2.4. Time-rock units

Ideally, each successively younger stratigraphic unit has successively fewer and morphologically sharper superposed craters. As with the dating of terrestrial sediments by fossils, this ideal correlation is not always attained. A formal stratigraphic practice has been developed for terrestrial geology to accommodate the distinction between the geologic units mapped on the basis of physical characteristics and the "clocks" that indicate elapsed time (North American Commission on Stratigraphic Nomenclature, 1983). Distinctions

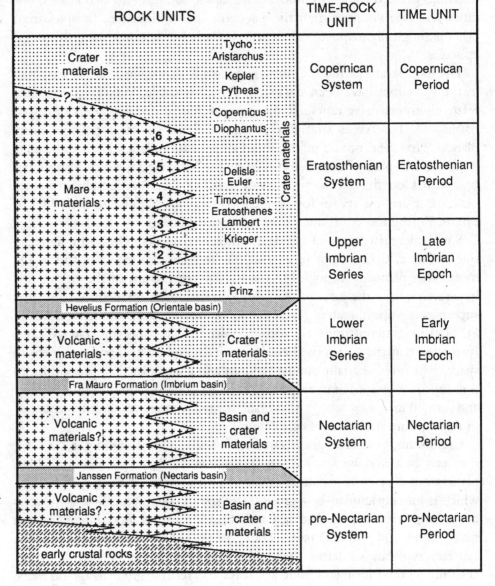

ROCK UNITS		TIME-ROCK UNIT	TIME UNIT
Crater materials — Tycho, Aristarchus, Kepler, Pytheas, Copernicus	Crater materials	Copernican System	Copernican Period
Mare materials — Diophantus, Delisle, Euler, Timocharis, Eratosthenes, Lambert, Krieger		Eratosthenian System	Eratosthenian Period
Mare materials — Prinz		Upper Imbrian Series	Late Imbrian Epoch
Hevelius Formation (Orientale basin)			
Volcanic materials	Crater materials	Lower Imbrian Series	Early Imbrian Epoch
Fra Mauro Formation (Imbrium basin)			
Volcanic materials?	Basin and crater materials	Nectarian System	Nectarian Period
Janssen Formation (Nectaris basin)			
Volcanic materials? / early crustal rocks	Basin and crater materials	pre-Nectarian System	pre-Nectarian Period

Figure 7.6
Lunar columnar section indicating interfingering rock units of crater materials and mare materials (examples from Mare Imbrium and Oceanus Procellarum; Wilhelms, 1980) and relations among rock, time-rock, and time units.

among *rock, time-rock,* and *time* units based on this Stratigraphic Code have benefitted planetary geology (Figure 7.6).*

Rock (rock-stratigraphic or lithostratigraphic) units are the three-dimensional physical units that may be directly observed and mapped. The fundamental rock unit is the *formation;* formations can be combined into *groups* and divided into *members*. For example, several individual deposits of the lunar Orientale basin are considered formations (Hevelius Formation, Montes Rook Formation, Maunder Formation), which collectively constitute the Orientale

* Although this 1983 edition of the Stratigraphic Code recommends the terms *lithostratigraphic, chronostratigraphic,* and *geochronologic,* the simpler terms are used here.

Group (Scott et al., 1977). In contrast, maps may treat each facies of an individual crater as a member, all the facies of each crater together as a formation, and several such formations of related craters as a group. Each formal rock unit is defined in a *type area* to which other occurrences can be compared.

Time-rock (time-stratigraphic or chronostratigraphic) units include all the rock units emplaced on a planet within a given timespan. Time-rock units consist of rock units and are defined on the basis of specific rock units in specific type areas. Thus, the time-rock unit is a physical unit. The basic time-rock unit is the *system,* which can be divided into *series* (and finer subdivisions in terrestrial geology). The deposits of the lunar Imbrium basin (more specifically, the Fra Mauro Formation) define the base of the Imbrian System, and the Orientale Group divides this system into lower and upper series (Wilhelms, 1987). The Lower Imbrian Series includes the materials of both basins and whatever other materials intervene stratigraphically on the whole Moon. Older systems are the pre-Nectarian (informal) and the Nectarian (Stuart-Alexander and Wilhelms, 1975). For want of a better criterion, the top of the Imbrian System and of the Upper Imbrian Series is defined by certain rock units of mare basalt, and geographically separated rock units are correlated by comparing their crater densities with those in the type areas. Younger systems are the Eratosthenian and Copernican (Shoemaker and Hackman, 1962). Time-rock units do not overlap; the upper boundary of one is the lower boundary of the next. This exclusivity is not always true of rock units, which commonly intergrade laterally (e.g., clastic and carbonate sediments on Earth, groups of lavas and crater deposits on other planets) (Figure 7.6).

Time (geochronologic) units are defined as the time during which a corresponding time-rock unit was deposited and are not physical units. *Periods* correspond to systems, and *epochs* to series (e.g., Imbrian Period, Early Imbrian Epoch).

These distinctions facilitate geologic mapping. Many rock units can be mapped from their distinctive textures or remote-sensing properties without knowing their ages. In other cases, relative ages of a terrain can be determined from crater frequencies before the individual rock units are fully delineated. These two types of information then converge when time-rock units are devised. The separation of rock and time-rock concepts also allows one or the other assignment to be changed without overthrowing the entire system of stratigraphic nomenclature.

The history of lunar mapping again provides illustrations. Although the stratigraphic scheme devised by Shoemaker and Hackman (1962) was conceptually powerful and generally very successful, its nomenclature implied too close a correspondence between rock and time-rock units. Shoemaker and Hackman (1962) concluded from telescopic crater counts that all lunar mare materials had about the same age as those that lie stratigraphically between the materials of Archimedes and Eratosthenes (Figure 7.4, *MI;* Figure 7.5). The rock unit "mare material" on the whole Moon was equated with the time-

223

rock unit "Procellarian System," which intervened between the "Archimedian Series" (mostly the Archimedes-type craters) and the Eratosthenian System. In fact, however, many mare units are older than Archimedes or younger than Eratosthenes. To accommodate this wide spread in mare ages, the Procellarian System was dropped, and mare units (and other lunar rock units) are assigned to whatever system their crater densities, crater morphologies, and stratigraphic relations indicate they belong (McCauley, 1967; Wilhelms, 1970).

Similar distinctions between rock and time-rock units can help avoid prejudgments about the histories of other planets. A scheme based on the lunar scheme has been devised for the geologically similar planet Mercury (Spudis, 1985). The two oldest systems are divided by the Goya Formation, the lineated peripheral material (probably ejecta) of the Tolstoj basin, into the pre-Tolstojan and Tolstojan Systems. These are parallel to the lunar pre-Nectarian and Nectarian Systems. The top of the Tolstojan System lies at the base of the Caloris Group (McCauley et al., 1981), which includes all materials of the Caloris basin. The Calorian System includes extensive light-colored plains, which may be either contemporaneous with or younger than the Caloris basin. The stratigraphic scheme flexibly allows for both possibilities by allowing the plains deposits to be mapped as rock units without requiring a decision about their exact stratigraphic position. The top of the Calorian System lies below the deposits of the crater Mansur. The two youngest systems are the Mansurian and Kuiperian, which, like the lunar Eratosthenian and Copernican, include craters that are slightly degraded and bright-rayed, respectively. Although morphology and rays are physical distinctions that are not entirely time dependent, they are a useful basis for approximate time-rock units.

Three systems organize the more complex Martian stratigraphic column (Scott and Carr, 1978). The oldest, the Noachian System, includes the relatively intact, ancient, basin- and crater-rich uplands including "cratered plateau material," which is basically undivided uplands material recalling the lunar "undivided terra material." The intermediate and younger systems, the Hesperian and Amazonian, respectively, include diverse plains deposits, extensive channel deposits, lobate flows of lava and other materials, probable and possible eolian deposits, and a great variety of other deposits that characterize this diverse planet. These three systems are defined on the basis of type areas of key units but are less well defined than those of the Moon and Mercury because extensive stratigraphic datum planes are rare on Mars. Most Martian rock units are therefore assigned to systems on the basis of crater densities, which are correlated with the crater densities measured in the type areas. Many groups of martian rock units cross the time-rock boundaries. Despite the approximate nature of the systems, they usefully provide a first-order classification for the complex martian rock-unit stratigraphy.

No formal time-rock schemes have been devised for the four highly diverse Galilean satellites of Jupiter (Morrison, 1982). All the identifiable stratigraphic datum planes on Io are local, and many are so young that they were created between the flybys of the two Voyager spacecraft. The relatively featureless

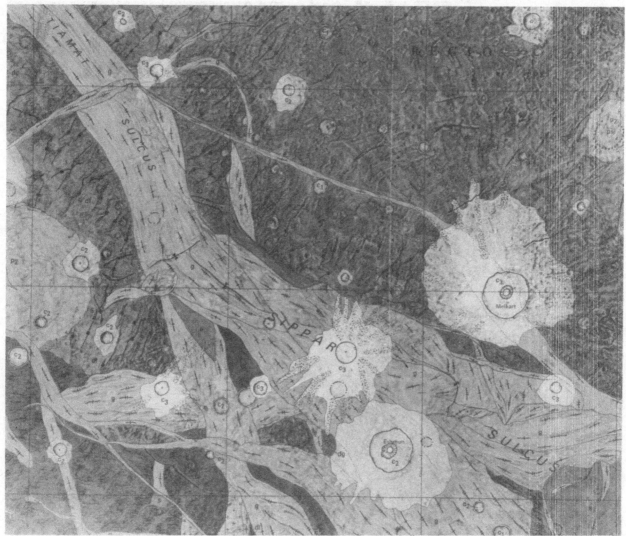

Figure 7.7
Geologic map of part of the Uruk Sulcus quadrangle of Ganymede, showing truncation of older, darker, indistinctly structured terrain by younger, brighter, closely grooved terrain. Several crater units also mapped. Air-brushed base; approximate boundaries 4°N–22°S, 180°–212° W. (Guest et al., 1988).

Europa consists of two basic units, extensive light-colored materials and streaks and patches of dark material. Descriptive rock-stratigraphic schemes probably suffice for these two satellites. The materials of Ganymede are divided into two basic groups that approximately correspond to time-rock units: a generally older, darker, lineated-to-hummocky group and a younger, brighter, grooved-to-smooth group (Figure 7.7). The ejecta and secondary craters of the basin Gilgamesh provide an excellent local stratigraphic reference plane that is a potential planetwide correlation reference for Ganymede. Callisto is characterized by multitudes of craters and by impact basins with more concentric rings than are known on any other planet. The larger of these basins are also potential stratigraphic markers.

The small satellites of Mars and Saturn display stratigraphic sequences of heavily cratered uplands and more lightly cratered units (Figure 7.8) (Smith

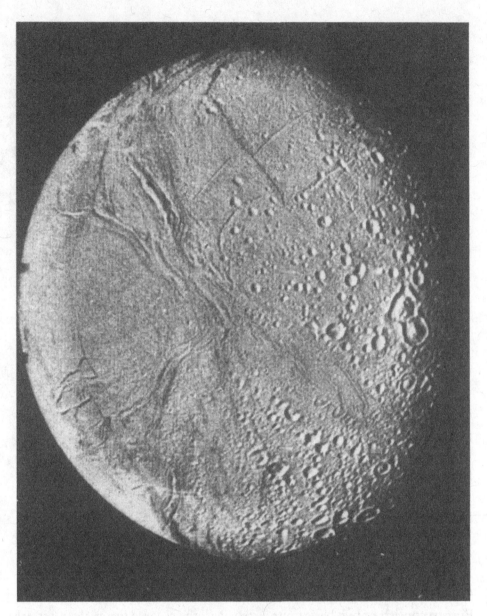

Figure 7.8
Saturnian satellite Enceladus,
showing a densely cratered unit
and younger smooth and structured units (Smith et al., 1982,
p. 520). This geologic diversity
was unexpected on such a small
body (502-km diameter). Voyager 2 frame 1715S2-001.

et al., 1982; Plescia, 1983) that could be formally recognized as time-rock units if it were considered useful. Local stratigraphic relations are abundantly clear in the Soviet radar images of Venus (see Chapter 2), and the future unraveling of the complex geology of this fascinating planet will probably benefit from the establishment of a time-rock scheme.

7.2.5. Structures and structural units

Tectonic structures modify material stratigraphic units on all planets – to the greatest extent on Earth, Venus, and Ganymede; the least on some of the small icy or rocky satellites. Investigation of the spatial and temporal relations of structures is an integral part of geologic mapping. Structures and structural patterns should be related to specific rock or time-rock units and therefore to the evolutionary history of the planet.

Linear features are often plotted on rose diagrams and the like for apparent objectivity. However, many such features are not true structures but are artifacts of lighting angle and other subjective factors. Features known to be true structures should be mapped where possible. Individual structures are mapped by line symbols adapted from terrestrial mapping (see Section 7.4.5). Treatment of extensive tectonically deformed terrains requires more attention here.

Because structures commonly cut across several rock units, the Stratigraphic Code does not allow for mapping of units on the basis of their structural modification alone. Where the rock units of a highly faulted or otherwise deformed terrain are recognizable, they should be mapped as separate rock units. Sometimes, however, the deformed rock units are not recognizable. In this case it is better to map such structural units as "fractured plains material" (Figure 7.9) than to ignore the presence of the structures in order to adhere strictly to the Code.

Some structures are closely tied with material units. Both of the major types of unit on Ganymede that are characterized mainly by tectonic patterns and albedo (Figure 7.7) probably consist of similar ice-rich materials. However, the younger, lighter, grooved and smooth units probably consist of new or recycled material that broke through the surfaces of the older, darker, lineated or hummocky units. Mapping these units and subdivisions by structural patterns conforms with the Stratigraphic Code because the material coincides with the structural pattern.

Martian chaotic terrain illustrates the question of material versus structural units and also the relation between mapping conventions and scale. The chaotic topography results from loss of coherence of other units. Large tracts of this terrain are visible on medium- and low-resolution images and are mappable at scales of 1:15,000,000 or 1:5,000,000 as chaotic material (Figure 7.10). Some blocks and block-bounding fractures that are included with the chaotic material at 1:15,000,000 scale can be mapped individually at 1:5,000,000 scale. The higher the resolution and the larger the map scale, the more numerous are the blocks of the chaotic material that can be mapped separately as "plateau material," "plains material," or whatever unit broke up chaotically.* Despite its origin by a quasi-structural process, the chaos may be considered a new material unit, because its source materials have been physically reconstituted at the scale of mapping. A terrestrial analogy is landslide breccia consisting of jostled blocks that are commonly derived from

* Scales are referred to here as "small" or "large" according to their numerical value. Thus, 1:15,000,000 is a smaller scale than 1:5,000,000. By this convention, a map at a small scale covers a larger area than a map at a large scale.

A

Figure 7.9
Heavily fractured and smooth
units in the Alba–Tharsis region
of Mars. A. Part of photomosaic
of the Tharsis NE quadrangle
(M 2M 22/101 CM, MC9 NE)
(U.S. Geological Survey; original
scale 1:2,000,000).
(Continued)

different units. In the final analysis, all terrestrial and planetary geologic units are recycled, to different degrees, from older units (see Section 7.2.1). Some recycling is physical; some, chemical.

7.2.6. The role of erosion

In mapping from images, one is faced with the necessity of distinguishing depositional from erosional surfaces. The distinction is particularly subtle in the case of undistinctive surfaces, such as plains and smooth uplands. If depositional, the exposed surface has the same age as the deposit. If erosional, the surface is younger than the unit or units that it exposes. Most of the outcrop patterns seen on Earth-satellite images were created by erosion, and the original depositional patterns of the stratigraphic units that compose these landforms must be reconstructed with allowance for the erosion.

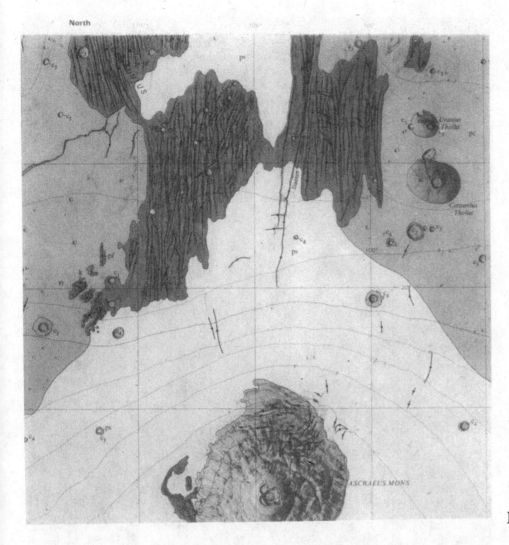

B

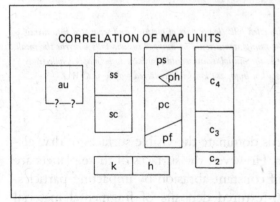

CORRELATION OF MAP UNITS

Figure 7.9 (cont.)
B. Geologic map including area of 7.9A and additional area to south, based on Mariner 9 data (Carr, 1975). Scale: each 10 degrees of Martian latitude covers about 600 km. C. Part of explanation of 7.9B, showing interpretations of stratigraphic relations; au, Olympus aureole material; c_4, c_3, c_2, crater materials, in order of interpreted increasing age; h, hilly material; k, knobby material; pc, pf, ph, and ps, cratered, fractured, hilly, and sparsely cratered plains materials, respectively; sc and ss, cratered and sparsely cratered shield material, respectively.

C

229

Figure 7.10

Martian chaotic terrain, broken up into small, jumbled hills (lower part of picture) and separated into miniplateaus (upper part). At the scale of this image, the miniplateaus could be mapped individually whereas the small hills could not; at smaller scales (lower resolutions), the miniplateaus would also have to be mapped collectively as chaotic terrain. Viking orbiter frame 366S61, 254 km high, 284 km wide, centered 0.9° S, 34.8° W.

In contrast, depositional patterns dominate the visible surfaces of dry, airless, tectonically quiescent planets. However, the surfaces of these planets are exposed to erosion in the form of constant abrasion by impacting particles. Most slopes are covered by soft-textured deposits of fragmental material chipped from the bedrock. On the Moon, some of this material has crept downslope and accumulated at the slope bottoms in molding-like deposits (Figure 7.11). This material therefore constitutes three-dimensional units superposed on the older units from which it was derived and even on units that postdate the older units (Figure 7.11C). In this role of forming new units created from old materials, the slope debris is like the debris eroded and transported by water, ice, and wind on more active planets, except that it has remained close to its source.

The one imaged planet besides Earth where erosion is known to have played a major role is Mars. The present map patterns of many Martian units have been created by erosion. Truncated edges of strata, evocative of terrestrial geology, are exposed in erosional scarps. Irregular reentrants in the map plan of many deposits probably indicate that erosion has stripped back the edges of the deposits and exposed an underlying layer (Figure 7.12). As on conventional terrestrial geologic maps, the uncovered unit and the remaining parts of the eroded unit are assigned the ages of their deposition, not of their erosion. The age of the erosion surfaces may be shown by supplementary conventions (Milton, 1975).

On many planets, textures produced by erosion are often more evident than textures produced by deposition. In layered sections on Earth and Mars, resistant beds protrude and weak strata recede. Horizontal surfaces are also differently affected by erosion. For example, the wind may scour a soft Martian deposit while leaving an adjacent deposit untouched. Although not intrinsic to the unit's deposition, these erosional properties are valuable indicators of the units' lithologies.

A mapper is not always sure whether textures and morphologies are depositional or erosional in origin. The striated walls and floors of Martian channels or valleys are clear indications of the flow of some fluid material (water, ice, lava). However, the striations may be grooves scoured by erosion, the outcropping edges of strata exposed by the flow, or depositional features of the sediment deposited by the flow. The striated surfaces can be mapped descriptively but cannot be fully integrated into the Martian stratigraphic scheme if their origin is unknown. The striations may affect more than one depositional unit. They should be objectively mapped and approximately dated pending further information. A similar example is the mapping as "gullied terrain" of Martian upland terrains that are characterized by intricate valley networks. These terrains may or may not coincide with stratigraphic units.

7.2.7. *Interplay of mapping and theory – a lunar case history*

Determining emplacement mechanisms is assessed at two levels in two stages of mapping. The first, already discussed in general terms, establishes that a rock body is a unit by recognizing that its appearance and stratigraphic relations are consistent with formation by a single general process (radial deposition from a central source, mantling by particulate material of varying thickness, viscous flow, flooding of an extended surface, fluidlike movement in a restricted channel, etc.). Associations between units are another clue obtainable by mapping; consistent juxtaposition in large areas or the whole planet suggests that two units are genetically related. Mapping will also show whether a feature is part of an extensive unit and therefore significant to the planetary geologic style or is merely a unique eye-catching anomaly that need not de-

A

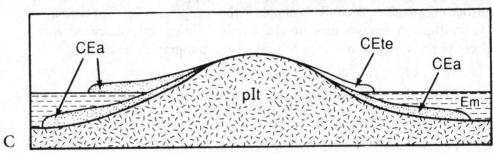

B

C

Figure 7.11
Part of premare lunar crater Flamsteed P, showing softened landforms and basal "moldings" resulting from
downslope movement of particulate material. View centered at 1.75° S, 43.2° W; largest crater is 1.7 km in
diameter. A. Lunar Orbiter 1 frame M-194. B. Geologic map (Offield, 1972). Debris forms Copernican or
Eratosthenian units CEa (apron material) and CEte (terrace material) at bases of near-exposures of pre-
Imbrian terra (crater) material (pIt). C. Diagrammatic cross section illustrating interfingering debris and
mare-basalt units. Units CEa and CEte are superposed on unit Em; thus, old but recently mobilized material
overlies relatively young unit. Earlier accumulations of unit CEa formed before, and are overlain by, unit Em.

Figure 7.12
*Irregular scarp on Mars (ar-
rows). Material between arrows
exposed by erosion of overlying
layers. Viking orbiter frame
369S32, 259 km high, 249 km
wide, centered at 6.3° N, 8.0°
W.*

mand much attention during early studies. Mapping requires attention to the
full geologic scene and not merely to special features that may give a partial
or false picture.

The second interpretive step is to infer more precisely what the process was
(primary or secondary impact cratering, pyroclastic volcanism, lava extrusion,

aeolian deposition or erosion, etc.). This refinement usually requires that the clues to emplacement processes discovered during mapping be supported by theory, experiment, field study of analogous terrestrial features, or actual samples of the unit or a similar unit. The history of the interpretation of lunar craters and of the peripheries of ringed impact basins is recounted here in some detail to illustrate this interaction between mapping and other studies.

A major contribution of geologic mapping is to show whether all parts of a given terrain formed simultaneously. Shoemaker (1962b) interpreted the crowding of satellitic craters around the crater Copernicus as a sign that they formed simultaneously with Copernicus (Figure 7.3A). He also analyzed their pattern in detail on the basis of ballistic theory and his field study of Meteor Crater, Arizona. He found that all observed features are consistent with origin of Copernicus by a primary impact of a cosmic body followed by secondary impacts of the ejected material onto the surrounding surface.

Next, Shoemaker and Hackman (1962) showed that secondary-crater swarms of Eratosthenes and Archimedes are similar to those of Copernicus, even though they are more subdued and rayless and those of Archimedes are, furthermore, cut off in some sectors by mare materials (Figure 7.4). Certain features of other complex craters, however, seemed to be inconsistent with the proposed simultaneous origin of all crater parts by impact. Some craters' central peaks seemed to have summit craters like those of volcanoes, thus to differ in origin and age from the rest of the crater. Superposition of smooth pools on some craters suggested, similarly, that the pools' material was post-impact and volcanic. The satellitic pits of Copernicus that compose the Stadius rilles (Rimae Stadius) seemed to be volcanic craters that were unrelated to Copernicus because of their nonradial alignment (Figure 7.3A, letter *S*).

These findings required either the development of endogenic (internal origin) hypotheses or refinements of the impact model. Some investigators concluded from these anomalies that the entire primary crater and its satellitic craters are endogenic. In the 1960s, however, hybrid theories were in greater favor; volcanism was thought to modify impact craters in diverse ways and to produce a wide variety of other lunar landforms as well.

Better data enabled the competing hypotheses to be tested. Lunar Orbiter images (1966–7) resolved the "volcanic pits" of central peaks into irregular depressions amid jagged parts of the peaks, like the depression in a molar tooth. Thus, these are typical lunar central peaks. Studies of terrestrial meteorite craters were concurrently demonstrating that central peaks originate by the immediate rebound of impact-crater floors. The Orbiter images also revealed that the Stadius chains, although not radial to Copernicus in overall map plan, have ejected material that *is* radially disposed. This observation removed the main argument against secondary-impact origin. In a classic case of laboratory support of an observation-based hypothesis, impact origin was conclusively demonstrated when experiments at the NASA Ames Research Center, California, reproduced the configuration of the craters and their ejecta down to the last detail by near-simultaneous impacts at different timings and

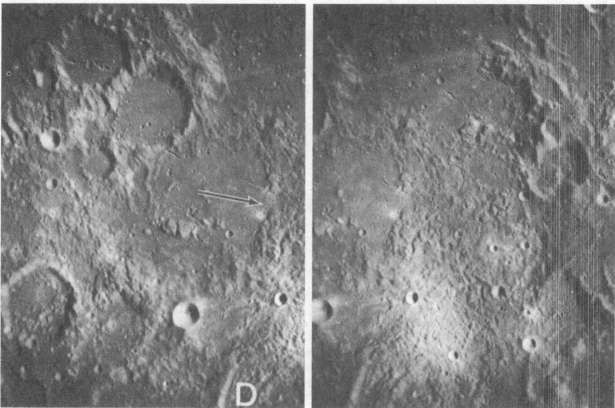

Figure 7.13
Region of Apollo 16 landing site showing changes of mapping caused by changes of interpretation. A. Stereo-
scopic photographs (see Figure 7.16). Arrow, landing site (9.0° S, 15.5° E); D, crater Descartes (48-km
diameter). Apollo 16 frames 974 (right) and 976 (left). (Continued)

spacings. The smooth pools were the last "postimpact, volcanic" part of craters to be reinterpreted in impact terms. Superior images of fresh lunar craters revealed flow textures between the pools and adjacent gradational veneers that indicate an extensive coating by some material. The material was identified as impact-melt rock by analogy with the melt rock that overlies large areas of some terrestrial impact craters.

These sophisticated interpretations were substantiated by samples of lunar material and, reciprocally, helped to explain the samples. Complex, multiply reworked impact breccias were found at every terra site visited by Apollo and the Soviet Luna unmanned samplers, including the Apollo 16 site, where volcanic rock had been predicted and sought (Figure 7.13).

As a result of these new data, geologists no longer had to equally weigh endogenic and impact hypotheses for the origins of geologic units and could reassess the features that had stumped them. Their mapping of the peripheries of impact basins had revealed diverse crater chains and clusters, grooves, ridges, and even domelike forms that seemed superposed on some of the craters (Figure 7.13B). All of these could be explained by diverse volcanic processes.

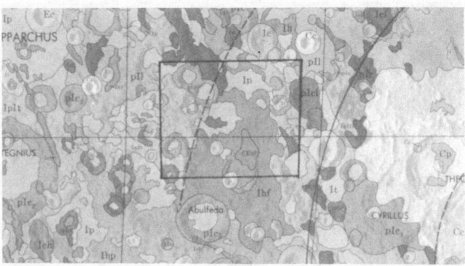

B

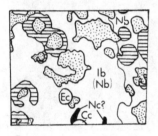

C

Figure 7.13 (cont.)

B. Geologic map prepared before the landing (Wilhelms and McCauley, 1971). Area of 7.13A outlined. Apollo 16 sampled units Ihf and Ip. C. Geologic sketch map of area of 7.12A incorporating post-Apollo results that indicate all units are of impact origin.

Previous (7.13B) and current (7.13C) mapping conventions and interpretations differ as follows. Units Cc₁, pIc₁, and pIc₂ (lower Copernican, upper pre-Imbrian, and middle pre-Imbrian crater materials, respectively) were and are interpreted as impact deposits and are not remapped. Units CEhf (Copernican or Eratosthenian hilly and furrowed material) and Cp (Copernican plains material) were believed volcanic; they are deleted because CEhf is reinterpreted as superficial ray material superposed on older hilly and pitted terrain, and Cp (outside remapped area) is probably impact-melt rock contemporaneous with unit Cc₁. Morphologically diverse units Ich (Imbrian crater chain and cluster material), Ici (Imbrian irregular-crater material), and pIci (pre-Imbrian irregular-crater material) were considered either secondaries of basins or volcanic; secondary origin is now favored (Sec. 7.2.7) for them and for some Ic and pIc, and all are combined in 7.13C as one unit (horizontal line pattern). Imbrian units Id, Ih, Ihf, Ihp, and Ip (dome, hilly, hilly and furrowed material, and plains materials, respectively) were interpreted as volcanic; unit Ip is retained (stipples) but is reinterpreted as impact-generated; the other units and unit pIl (thought to be pre-Imbrian material pervasively faulted during Imbrium impact) are combined as Imbrium-basin material (unit Ib). Unit It (undivided Imbrian terra material) was considered of uncertain origin; it is now considered a minor surficial cover and is deleted. Unit IpIt (undivided Imbrian or pre-Imbrian terra material) was considered to be basically of basin origin, and unit pIr (pre-Imbrian rugged material) to be pre-basin rock uplifted by impact; both units are now combined as Nectarian basin material (unit Nb). Crater Descartes (former unit pIc₁ of uncertain origin) is given the equivalent current designation Nc?

However, reexamination revealed previously unsuspected similarities in morphology, map pattern, distance from the source, and secondary-to-primary size ratio between these large features and secondary craters of Copernicus and the laboratory models. Therefore, the circumbasin craters and related landforms were probably created by secondary impacts of basin ejecta. Because all impact structures follow the same general plan, even subtle, isolated landforms could now be traced to their source basins and used to date the units that they touch (Figure 7.13C).

This new knowledge was especially useful in the interpretation of nondistinctive units that had been set aside during early mapping as "undivided terra

Figure 7.14
Lunar crater Tycho (85-km di-
ameter; 43.3° S, 11.2° W), sur-
rounded by finely textured ejecta
and secondary craters (compare
Figure 7.2). Other craters in
view lack such sharp detail. Lu-
nar Orbiter 4 frame H-119.

material" (units It and IpIt, Figure 7.13B). These rocks could now be rein-terpreted by an application of the principle of uniformity. One can tentatively assume that the nondistinctive terrain formed in the same way as the better-expressed terrain of a younger analog. For example, the lunar crater Tycho possesses, but nearby craters lack, the diagnostic indicators of impact origin (Figure 7.14). Because of the proven dominance of impact on the Moon, one can reasonably assume that older craters also once had distinctive radial ejecta and secondary craters like those of Tycho, but that the old craters have been eroded by the ubiquitous rain of smaller impacts that has always attacked the Moon at all scales (Section 7.2.3). Degraded basins, similarly, once resembled fresh-appearing basins like Orientale.

The new understanding of the Moon has greatly simplified lunar geologic mapping and models of lunar evolution. When every peculiar landform was thought possibly to be a volcanic extrusion of different composition or age and when each crater might have a different origin, mappers did not know what was significant and what could be omitted. As a result, geologic maps were complex. Now, a geologic map of the lunar terrae can consist, essentially, of the overlapping deposits of some forty basins.

In summary, the steps toward the current understanding of the Moon included (1) observations based on images (satellitic craters of Copernicus), (2) theory (ballistics), (3) multiplication of hypotheses to cover apparent anomalies (the "postcrater" features), (4) acquisition of new data (images and samples obtained by spacecraft), (5) laboratory testing (simultaneous impacts) and geologic field study of analogs (terrestrial impact melts), and (6) testing and application of the new paradigm by renewed geologic mapping. In my opinion, the new impact paradigm has not been found faulty, but it is subject to further testing and revision with new data.

Martian studies cannot yet be generalized in this way because many endogenic features are present, a variety of partially understood erosional and depositional processes have operated, and the uniformitarian principle has more limited validity on complex planets than on airless, waterless, relatively simple planets. Martian geologic mapping, therefore, is still relatively complex. Nevertheless understanding of Martian geology is advancing rapidly because of the good coverage by images and maps (see Section 7.3) and because Mars can be compared to and contrasted with both the Earth and the Moon.

This account has shown that neither geologic mapping nor any other kind of study is sufficient by itself. As the literature clearly shows, theory that is not based on examination of actual planetary surfaces is as unrealistic as it is in terrestrial geoscience. Similarly, mappers require nonmapping data to confirm that their genetic interpretations are physically possible and to help them choose between multiple working hypotheses.

7.3.   DEALING WITH PLANETARY IMAGES

Although a complete job of planetary geologic mapping requires nonvisual "remote sensing" data of several types (see Section 7.4.1; Basaltic Volcanism Study Project, 1981, Chap. 2) and, eventually, samples collected from the surface, the richest data sources for the mapping are images. Experience has revealed several difficulties in viewing images that are encountered by every beginner.

The first is seeing the relief reversed – depressions as elevations and elevations as depressions. Illumination from the upper left of a scene apparently causes the fewest problems, but this orientation is not always achievable. I prefer always to keep north at the top, to memorize best the appearance of scenes. The reversed-relief annoyance can be overcome by training one's perceptions to react to certain clues. For example, the most abundant circular forms on impact-dominated surfaces are craters, that is, depressions. Martian surfaces often present greater difficulties, and even experienced observers have trouble with this morphologically diverse planet. Here, too, impact craters can be found and imprinted on the mind as negative. Other landforms will then seem to fall into their correct relief.

Rational analysis of sun-illumination direction is often needed in addition. Slopes facing the sun appear to be much brighter than those facing away from it, and those facing away may be in shadow (Figure 7.2A). This analysis may have to continue throughout a study of Mars, because the Mariner and Viking orbiter images are more diverse in sun illumination than are Lunar Orbiter images (most of which were taken at rising-sun illumination on the nearside and at setting-sun illumination on the farside). The need for caution is illustrated by the fact that some Martian mosaics include frames with opposite illuminations; what appears to be a trough or crater on one frame may appear to change into a ridge or dome on the next. The flyby missions that imaged the other planets produced more uniform illumination directions.

The angle between the sun's rays and the surface is another variable. The same feature looks very different at low and high solar illuminations (Figures 7.3, 7.15). Low-angle illumination emphasizes topography, and high-angle illumination emphasizes brightness differences of features like crater rays and lunar maria. An area of lunar mare may look so rough at grazing illumination near the terminator (the boundary between shadowed and illuminated terrain) that it does not look planar (Figure 7.15B). Differences may be evident on the same or different images. In a common example, the sun-facing slope of a given crater is so bright as to appear washed out, whereas the opposite slope of the same crater, on the same image, shows considerable topographic texture because of grazing sun illumination. Although the long shadows on low-sun images usefully bring out subtle topographic features, the shadows may completely obscure the insides of craters (Figures 7.2A, 7.15B). Images taken at two or more illuminations are therefore needed to fully interpret most geologic units. In comparing units, the observer must either compare images taken at about the same illumination angle or compensate mentally for the differences. Unawareness of the effects of lighting once led naive observers to interpret long shadows cast by boulders as missiles or spirelike monuments. For lunar photogeology, sun-elevation angles of about 1 to 20 degrees for smooth surfaces and 10 to 45 degrees for rugged surfaces have been found to be optimal. The availability of controlled stereoscopy for some images (Figures 7.13, 7.16) has not diminished the importance of sun angle.

Another problem is adjustment to varying resolution of the images. Resolution is defined as the size of an object that can be identified (identification resolution) or detected (detection resolution). Planetary images vary widely in resolution, and the different resolutions have different uses. Large-scale, detailed images are usually necessary to show the textures and contact relations upon which most interpretations of origin are based. Small-scale, regional images are useful for mapping the distributions and associations of units, which are also valuable in interpretations, as discussed in the previous section. For the Moon and similar planets, the large-scale images show details of the

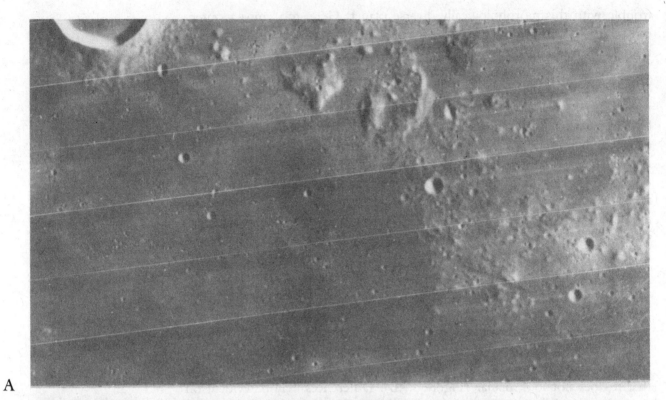

A

B

Figure 7.15
A lunar area under different sun illuminations. Large crater at top is Kunowsky (18-km diameter; 3.2° N, 32.5° W). A. Sun angle averages 18.5 degrees above horizontal. Lunar Orbiter 4 frame H-133. B. Sun angle from 3 degrees above horizontal at right to grazing at left. Apollo 14 frames 10375-10377 (right to left).

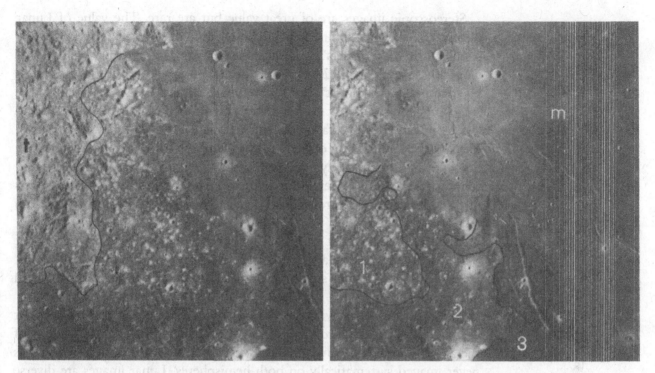

Figure 7.16
Stereoscopic pair of photographs of region along the contact between Mare Serenitatis (m) and Montes Haemus (compare Figure 7.4). For a three-dimensional view without a stereoscope, look at a point on the left photograph with your left eye and at the same point on the right photograph with your right eye, let your eyes unfocus, cross them until the points converge, and then refocus. Dark lunar blanketing material of variable thickness may be mapped, alternatively, everywhere it is visible (are as 1–3), only where it is most conspicuous (areas 2 and 3), only where it completely covers the terra (area 3), or not at all. t, terra material that is probably exposed (left frame). Apollo 17 mapping-camera frames 2102 (right) and 2103 (left); each frame covers an area 300 km high.

regolith that covers most of the surface, whereas the small-scale images show the configurations of the basic bedrock units – those of impact basins and maria. For Mars, images that cover very large areas are of little value for geologic purposes. Medium-resolution views (taken at ranges of about 6000 to 10,000 km above the surface) show the major geologic units. Higher-resolution Martian views potentially can reveal details of stratigraphic relations and textures that are diagnostic of the units' emplacement and modification processes. In particular, such views may reveal geologic units of the same sizes as terrestrial geologic units, which are not resolved at the regional scales.

Ideally, low- and high-resolution images are available of the same area and are used in tandem. Information from small points viewed at high resolutions can then be extrapolated to the larger areas. Subtle properties visible on the small-scale images often assume new significance after being seen in the magnified views. Section 7.4.1 shows how geologic mapping can be used to keep track of the different kinds of information obtainable from different kinds of images.

Stereoscopic images are of great value but are rare. The value of Lunar Orbiter stereoscopy is limited by the construction of each image out of narrow parallel strips ("framelets"), producing a stairstep effect when seen stereoscopically (Mutch, 1970). Most Apollo orbital images (true photographs taken by "mapping" or "metric" and "panoramic" cameras) provide excellent stereoscopy (Figures 7.13, 7.16) but cover only 20 percent of the Moon – and less than 20 percent at favorable sun illuminations (Masursky et al., 1978). The Mariner and Viking orbiters of Mars and the flybys of the other planets flew too high to routinely provide the base–height ratio necessary for good stereoscopy, although a few pairs do provide stereoscopic views. Direct views of the third dimension are at least as valuable for geologic purposes as high resolutions (Figure 7.16).

Discussion of the radar images of Venus is beyond the scope of this chapter. Suffice it to say that what appear to be shadows and bright slopes do not necessarily have the same significance as those visible on images taken in visual wavelengths.

The problems of viewing images can be overcome with experience, but even the most experienced observer may be troubled by the great diversity of the accumulated data bank. This is especially true for the Moon, which was never imaged systematically on both hemispheres. Lunar images are diverse in resolution, viewing angle (obliquity), format, and contrast (Mutch, 1970). The best Martian dataset, the Viking orbiter images, was acquired more systematically than the lunar images (Carr, 1981). Also, Mars is uniformly covered by excellent maps and photomosaics, particularly the very useful mosaics at a scale of 1:2,000,000 (see Chapter 3). The images of the other planets also vary in illumination, obliquity, and resolution but were acquired relatively systematically and are being mosaicked, catalogued, and rendered into maps so well that adjustment to their vagaries is not difficult.

7.4. MAP CONVENTIONS, FORMAT, AND PRODUCTION MECHANICS

7.4.1. General mapping procedure

The first step in examining a new planet or region is a general reconnaissance. The mapper scans the images and base map to gain a general idea of what kinds of units and structures are present, that is, the geologic style. This reconnaissance may take the form of a sketch map on a paper copy of the base map. A rapidly drawn sketch will focus the mapper's attention right from the beginning on the "big picture" – the geologic context and mutual relations of the most important units. Detailed examination of small areas and final interpretation of peculiar-appearing features can come later, after their possible significance has been assessed.

Planetary geologic mapping, like all science, progresses by building on previous work (see Section 7.2.7). It proceeds best by testing some working hypothesis or multiple working hypotheses, no matter how wrong they may later prove to be; purely inductive mapping is seldom productive. A mapper's success in this first reconnaissance will therefore depend on experience. The nongeologist will have no idea what to look for. The expert in the planet under study will look for familiar features and contact relations and also for those that appear new. The expert in another planet will have an advantage if the old and new planets have some common features and, furthermore, will have polished many mechanical mapping techniques that apply to all planets.

The purpose of the map should be decided early, because maps for different purposes are constructed differently. The most important decisions are scale and level of detail. There may be no choice about the scale if only one scale of base map is available, but a wide choice of level of detail is open even in this case. Some maps show the maximum detail that can be drawn and portray each contact faithfully (e.g., Figure 7.11B). At the opposite extreme are "cartoons" that show only the general relations among the most significant units or groups of units (e.g., Figure 7.17). Typically, a region is mapped in detail at large scales if it is poorly understood and is cartooned at small scales if it is well understood. In making the decision about degree of fidelity or cartooning, remember that a detailed map can be converted into a cartoon but not the reverse.

Whereas most geologic maps portray geology as it is today, special *paleogeologic* maps are sometimes constructed. Some paleogeologic maps portray only the geologic units that formed during some specific period or epoch. Others portray the cumulative record through that specific time (Figure 7.17B). Paleogeologic maps are usually derived from maps that show the present geology (Figure 7.17A).

Although most maps are constructed for eventual publication, some are constructed for personal purposes because of mapping's power as a learning tool. A mapper benefits from having to examine an entire area closely and from constant testing of hypotheses about unit origin and age against map relations and distribution. Although "learning maps" can bypass some of the guidelines described in the rest of this chapter, completion of a map through to publication is good training for all mapping.

I suggest that the best first step in actual mapping is to start in an area where familiar units are clearly expressed and to map their limits and relations. This process might be repeated in several parts of a large area or a whole planet. Temporary blank spaces are left in the less-well-understood intervening terrain.

A corollary of this progress from the known to the unknown is to start with the youngest unit in a region. Young units are usually the best expressed and most easily interpreted, and their contacts are the most complete and easiest to map. Terrestrial geologists commonly map the Quaternary alluvium first in order to focus on the diminished remaining area that contains the more significant and difficult problems. Similarly, planetary geologists com-

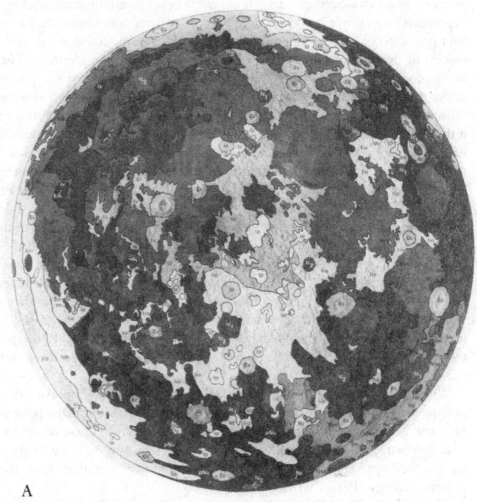

A

Figure 7.17

Simplified geologic maps of the lunar nearside (in Carr, 1984, pp. 197–205). A. Present Moon. Time-rock units (capitals): C, Copernican; E, Eratosthenian; I, Imbrian; N, Nectarian; pN, pre-Nectarian. Rock units (lowercase letters): b, basin material; c, crater material; d, dark-mantling material; i (second letter), Imbrium-basin material; i (third letter), inner deposits; m, mare material; n, Nectaris-basin material; o (second letter), Orientale-basin material; o (third letter), outer deposits; p, plains material; t, terra-mantling material; u, undivided material; v, volcanic-dome complexes. (Continued)

monly first map the young craters and well-defined patches of young plains. The contacts of young units truncate older contacts.

On Earth, units are often mapped by "walking" their contacts. One does the same, figuratively, in planetary mapping. Each unit is scanned until its end; then a contact is drawn. The contact is then traced until the two units it delimits no longer meet, whereupon it either terminates or continues as the contact between the first unit and a different second unit. If each unit with distinct contacts has properties indicating a common age and formative process over its whole area, and if its contact geometry and crater densities both

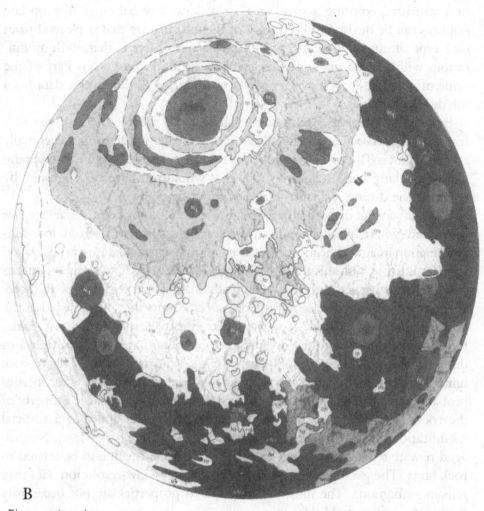

B

Figure 7.17 (cont.)
B. Paleogeologic map based on 7.17A, showing units that formed before and during the Lower Imbrian Epoch.

indicate the same age relations with adjacent units, all is well. Inconsistencies are noted if they appear. The process of transferring complex information from an image to the map in the form of contact lines becomes increasingly automatic with increasing experience.

A preliminary stratigraphic column should also be devised early in the project and updated throughout the study. Waiting until the end of a study to construct the column and the box explanation (see Section 7.4.6) inevitably causes unnecessary reworking of the geology.

Mappers initially will probably refer to the best images of large regions, but eventually they will examine all the available images. This may involve considerable struggle with the coverage. No planet and very few large areas of planets are favored by continuous good images (see Section 7.3). While poring through the image collection, the mapper can record whatever units or structures are seen clearly on each image by drawing a piece of a contact or a structure, coloring a small area, or making a verbal note. Appropriate symbols can be devised to keep track of the information that is gleaned from each type of image of an area – perhaps a series of letters that, with modification, will become geologic unit symbols (see Section 7.4.4). Part of the value of geologic mapping is its power to integrate all the diverse data from the different images.

Even if the mapper is temporarily ignorant of the geologic context of a feature glimpsed on an image, all the pieces of the puzzle should eventually come together. The overall geologic picture that emerges from this mapping and note-taking will probably be one that would not have been grasped by scanning the diverse available images without mapping.

Remote sensing data on albedo (Figure 7.3B), color spectra, radar, etc. are integrated where possible with the information obtained from low-sun-illumination images. The most easily used remote sensing data are in an imaged format (Basaltic Volcanism Study Project, 1981, Ch. 2). The imaged patterns can often be matched directly with already mapped geologic units. Remote-sensing data sometimes suggest the presence of units that were missed in the geologic mapping. "Color" (spectral reflectance) patterns of the lunar maria, for example, have suggested the presence of units that were later found to differ in density of superposed craters (e.g., Wilhelms, 1980). Where rock units and remote-sensing patterns coincide, the remote-sensing properties probably pertain to chemical composition or some other intrinsic property of the rock unit. In other cases, the remote-sensing data apply to a surficial modification that crosses unit boundaries (e.g., dust layers on Mars). Nonimaged remote-sensing data from a dense array of points can also be related to rock units. The geologic significance of scattered or low-resolution data may remain ambiguous. The most easily measured properties are not necessarily the most significant (Mutch, 1970, p. 58).

7.4.2. *Separation of interpretation from observation*

Each stage of mapping requires some interpretation. The act of drawing contacts involves the interpretation that three-dimensional geologic units are present. The criteria for defining the units usually have to be selected from among many possible sets of attributes. The significance, origin, and age of geologic units are inferred at several levels of precision, and these hypotheses may change during or after the mapping (Section 7.4.7). For the map to be credible and useful to the reader, therefore, the basis for all unit assignments and genetic interpretations must be made clear. In particular, the role of interpretation must be specified and distinguished from the objective data on which it is based. Several conventions for conveying this distinction have been developed.

One fundamental requirement in planetary mapping is the clear separation, in the explanation and text (Section 7.4.6), of the interpretation(s) of each unit from the list of objective physical characteristics by which the unit was identified by the mapper and can be reproducibly mapped. For example, one could distinguish the following:

Characteristics: Forms dark, extensive, level, mostly smooth surfaces having sharp contacts with adjacent terrain.

Interpretation: Basaltic lava

Second, units are given objective, not interpretive, names: crater rim materials or mare materials, not impact ejecta or basalt. During early lunar studies, when the origin of most craters was uncertain, the rim material might be interpreted alternatively as debris ejected from an impact crater, lava ejected from a volcanic crater, or the precollapse edifice of a caldera. Nevertheless, the unit could be mapped geologically and ranked stratigraphically, to a first approximation, on the basis of observational criteria. Even when the impact origin of most lunar craters was well established, the term "ejecta" was avoided because the radial deposits might contain more material dislodged by secondary impacts than primary ejecta from the crater itself. "Rim material" or "radial rim material" thus remained preferable to "ejecta" as descriptive terms. Similarly, an undistinctive planar deposit could equally well have been emplaced by volcanism, impact, or fluidlike sedimentation of debris, and each material can vary greatly in such properties as composition, viscosity, and grain size. In fact, the origin of some well-mapped and -dated lunar plains materials is still uncertain. The undecided mapper should state the uncertainties and give some alternative interpretations that are consistent with observation.

Placement of contacts should also be reproducible. A user of the map should be able to locate a contact on an image after having read the description of the adjacent units. However, the portrayal of units may differ among mappers who agree about interpretations. A thin surficial unit or a buried but still detectable bedrock unit may be mapped, depending on the map's purpose (see the following section). The exact placement of contacts may differ even among mappers who agree about which layer to map (Guest and Greeley, 1977, p. 17). In general, these differences are unimportant so long as the mapping is reproducible. Attempts to eliminate personal bias by quantifying mapping have proved unsuccessful because of the complexity of geologic terrains.

Each unit should be mapped in such a way that only its interpretation and not its contacts or status as a unit will have to be revised significantly when new data are obtained. When Apollo photography showed that the Serenitatis border unit in Figure 7.5 is older than the central unit, the symbols Em and Im could simply be replaced by appropriate new symbols without remapping or redefining any units. The finding by Apollo 16 that the plains in the Descartes region are of impact and not volcanic origin required changing only the interpretation and not the contacts, unit name, or time-rock assignment

247

of the plains unit (Figure 7.13, unit Ip). Changes of name, age, and interpretation, but not of contacts, were required when three units of crater material and parts of two more were reinterpreted as secondary craters of the Imbrium basin (Figure 7.13C, horizontal line pattern).

Sometimes, however, new data may require more fundamental changes. The hilly and furrowed Descartes Mountains were mapped as the rock unit Descartes Formation on the assumption that the distinctive furrows were intrinsic to the deposition of the unit (Figure 7.13). Extensive adjacent tracts were mapped as hilly and pitted material, and both units were interpreted as volcanic (Wilhelms and McCauley, 1971). When the Apollo 16 samples showed that the Descartes Formation consists of impact breccia, the furrows and pits were reinterpreted as secondary-impact craters superposed on earlier impact deposits. In this still-tentative interpretation, the craters are a modification and not part of the true depositional units. The secondary-crater deposit could now be shown to overlie Nectaris-basin material, Nectarian crater materials, or other underlying units (Figure 7.13C). Many of these new contacts do not completely coincide with those of the previously mapped units. Other interpretations might alter the contacts in other ways.

Many pitfalls have been avoided by sharply distinguishing interpretation from observation on planetary geologic maps. Nevertheless, the distinction has been weakened in recent years. Lunar mare material is now commonly called "mare basalt" – a reasonable change in view of the results of the lunar sampling. However, many mappers, especially novices, tend to weaken the observation–interpretation distinction for less securely interpretable units such as those of plains or small hills on Mars that seem more easily explained by volcanism than by less familiar processes. Unsuspected novelties may appear even in seemingly secure cases. Although the observation of currently active volcanism on Io, for instance, would seem to establish the origin of Io's units beyond all doubt, it does not demonstrate whether the present form of many units is depositional or erosional in origin or whether inactive depressions are vents or are calderas that never vented any material. Even when sampling or other on-site exploration establishes the origin of a unit, the findings might not apply to apparently similar units elsewhere. Readers of geologic maps with unsupported or unqualified genetic statements will be unaware that doubt exists. The mapper should state the basis for various possibilities as well as the reasons for favoring one.

7.4.3. Map units

At some phase of the mapping, preferably early on, the observed rock units are converted by some convention into map units – the units that are actually shown on the map and given a specific symbol, color, and position in the columnar map explanation. This conversion commonly involves recasting the observed units to make the map simpler or more readable or to remove some fallacy, such as confusion of rock and time concepts. Map units may be rock or time-rock units, of any rank, depending on the scale and purpose of the map.

As in all geologic mapping, planetary geologists must decide whether to lump or split when devising map units. Lumping is more common because more individual units are usually mapped in the early stages of a study than when the mapped area becomes better understood (Shoemaker, 1962a, p. 123). Splitting at late stages usually requires some remapping.

One extreme form of lumping is to show only the time-rock units. Another is to devise *"provinces"* that include diverse though related units. The main lunar provinces are the maria and the regions covered by basin materials of a given age. Martian provinces might be the Tharsis bulge, northern lowlands, cratered southern uplands, and channel-and-valley province (Carr, 1984, p. 211). These kinds of lumping are appropriate for small-scale synoptic maps, such as figures in books.

Some generalized map units combine diverse rock units whose age and origin are poorly known. An example is undivided terra material, which was called pre-Imbrian or Imbrian terra material (symbol, IpIt) on many lunar maps (Sections 7.2.2, 7.2.4, 7.2.7; Figure 7.13B). Devising one or two such "wastebasket" map units is a means of delineating unsolved problems. Unit IpIt proved to consist mostly of impact-basin deposits, which are now explicitly identified by the map units (Figure 7.13C, unit Nb).

A common type of lumping on planetary maps at all scales is to combine into generalized map units all the individual deposits that are physically similar and that are of about the same age (e.g., Copernican crater materials, Eratosthenian mare materials, or Imbrian plains materials). The materials may belong to the same time-rock system or series or, in favorable circumstances, can be stratigraphically bracketed more closely. The individual occurrences may differ in age within these limits, as do those of craters, or they may be contemporaneous, as are many separated patches of circumbasin lunar plains. Contrary to terrestrial practice, individual adjacent patches of the same map unit may be separated by contacts, the younger shown overlapping the older (e.g., Figure 7.5A, Aristillus and Autolycus).

Whether multiply occurring rock units are lumped or separately portrayed depends largely on their number relative to the map scale and on the availability of color for the final publication. Extensive rock units such as those of large impact basins may correspond directly with the map units. This treatment more nearly approaches terrestrial usage than does the lumping of many isolated crater or plains deposits into a single unit. Small-scale maps, however, may include the deposits of so many basins that lumping of all those of a certain age is desirable (Figure 7.17).

Fine distinctions made during mapping are sometimes retained. Crater-material subunits, for instance, once played a large role on lunar geologic maps (Wilhelms, 1970, pp. 40–2; Mutch, 1970, pp. 165–74). When little was known about lunar craters, objectivity required that rim, wall, peak, and floor materials be subdivided because they differ morphologically. Each could be interpreted according to the current state of the mapper's knowledge. Mappers could separate the parts of the crater they believed to be endogenic – as many parts were thought to be (Section 7.2.7) – from the parts they believed were of impact origin. The colored boxes for the units representing these parts in the map explanation (see Section 7.4.6) could show a substantial difference in age when arranged vertically and a slight difference when placed side by side. Because almost all parts of impact craters now appear to have been formed essentially simultaneously by the original impact, the age differences are considered minor and the parts ordinarily are not distinguished on maps. Remaining exceptions requiring separate map units might include unusual craters, craters mapped to investigate the cratering process, and floor-filling materials or other units that are still considered to postdate the rest of the crater.

Two or more superposed units are commonly recognized in the same area (see Section 7.2.2). The mapper then decides whether to show the surficial deposit, the underlying deposit that can be seen through the surficial cover, or both (Figures 7.13, 7.16, 7.18).A rule of thumb is to show the deposit that is topographically most conspicuous at the scale of the mapping. This unit will be the colored map unit. That is, the most conspicuous unit is shown by the most conspicuous convention. However, a map emphasizing basins will omit surficial units such as pyroclastic dark-mantling material (Figure 7.16) and a map emphasizing dark-mantling materials will probably omit the basins.

Conventions have been developed to show more than one superposed unit if desirable. Buried units whose textures are still visible may be shown by dotted contacts and symbols in parentheses (Figure 7.18B). Examples include basin materials buried by dark-mantling materials, plains contacts beneath crater deposits, and buried basin massifs. Where possible, buried contacts are drawn at the limit of their observed topographic expression rather than at the inferred or projected limits. An alternative is to show the underlying unit in color and the overlying unit by an overprint pattern, such as stipples or hachures (Figure 7.18B). This convention is useful for thin parts of dark-mantling materials or for rays of young craters. A third convention employed on some maps is to define a colored map unit as including underlying and overlying units. For example, the lunar map unit Imbrian and pre-Imbrian terra material commonly designated a pre-Imbrian crater or basin unit that was inferred to be overlain by a thin cover of Imbrian material. This designation differs from the already discussed map unit called Imbrian *or* pre-Imbrian terra material, whose age is not known more exactly.

7.4.4. *Unit names, letter symbols, and colors*

Each map unit is given a distinctive name, letter symbol, and color or pattern. Names may be formal or informal, as convenient. The U.S. Geological Survey developed many new formal names for lunar geologic units during the 1960s,

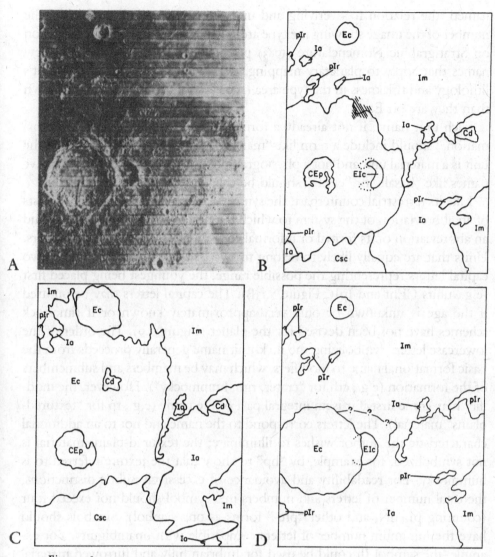

Figure 7.18
Different, equally valid methods of mapping an area. A. Photograph of Aristoteles region (crater 87 km in diameter; 50.2° N, 17.4° E). Lunar Orbiter 4 frame H-98. B. Most conspicuous units stressed. C. Surficial units stressed. D. Deep-lying units stressed.

but current practice favors informal names (crater material, mare material, ridged plains material, etc.). Formal names are still bestowed on rock units that are stratigraphically distinctive, laterally continuous, frequently discussed, or difficult to describe briefly. For example, the name "Medusae Fossae Formation" concisely and objectively describes an enigmatic, stratigraphically complex, morphologically diverse set of Martian deposits whose properties could not be expressed by a simple descriptive name.

Introduction of a new formal stratigraphic name requires a formal definition. Each definition includes a statement of intention to define a new name, the coordinates of the unit's type area, the feature after which the unit is named, the relation to overlying and underlying units, and, preferably, the number of the image showing the type area. The North American Commission on Stratigraphic Nomenclature (1983) provides guidelines for defining new names that apply to planetary mapping, except that such matters as a unit's lithology and thickness in the type area (type section) may be less well known than they are on Earth.

Each unit name, if not already a formal name including a term like "formation," should include a term like "material" or "deposits" to show that the unit is a material unit and not a physiographic form. As discussed, interpretive names like "basalt" or "ejecta" should be used very rarely and cautiously.

Like its terrestrial counterpart, the symbol for a planetary map unit consists of an abbreviation of the system to which the unit is assigned, in capitals, and an abbreviation of its formal or informal rock-unit name, in lowercase letters. Units that are equally likely to belong to two or three systems are given two capital letters representing the possible range, the youngest being placed first (e.g., units CEhf and IpIt, Figure 7.13B). The capital letters may be omitted if the age is unknown or only very approximately known or if time-rock schemes have not been devised for the planet (Figure 7.9). The order of the lowercase letters symbolizing the rock-unit name generally proceeds from the basic formational name to modifiers, which may be members and submembers of the formation (e.g., crh for "crater, rim, hummocky"). However, the modifier may come first if it is an integral part of the name (e.g., tp for "textured-plains" material). The letters correspond to the name and not to an additional characteristic the author wishes to illuminate; the textured-plains material is not symbolized, for example, by "hp" to show that the texture referred to is hummocky. For readability and avoidance of excessively subtle distinctions, the total number of letters and numbers in a symbol should not exceed four (counting pI, pN, and other "pre-" forms as one symbol). Symbols should have the minimum number of letters compatible with unambiguity. For example, the symbol Ih could be used for Imbrian hilly and furrowed material as well as for the Hevelius Formation; but a third letter must be added to one of the symbols if both units appear on the same map.

Where units are numbered, the oldest unit of a class is designated 1 (on line or subscript), and higher numbers refer to younger units (Figures 7.9, 7.13). Numbers follow all the letters, because they refer to the whole unit symbol (e.g., Icr_1, not Ic_1r). Numbers are undesirable as unit designators except for age distinctions because names and the corresponding letter symbols provide more direct clues to the units' properties.

In texts, a unit is referred to by its name or by the term "unit" plus the letter symbol, rather than by the letter symbol alone (e.g., "younger mare material" or "unit Im_2," rather than simply "Im_2").

The reason for querying a symbol on a map should always be given. A symbol like "Ec?" obviously means that there is some doubt about the assignment; the explanation should state whether the doubt is that the crater is younger, older, or a crater after all. Queries should be used sparingly because each one must be drafted on the final map.

Colors are chosen to associate like units and disassociate unlike units. Intense colors are used for small patches, weaker colors for extensive units. The best color schemes convey both rock-unit and time-rock associations. The readability of a map depends strongly on the choice of colors. Ideally, all maps of a given map series have similar schemes.

7.4.5. *Line symbols*

Line symbols on planetary maps follow terrestrial precedent as far as possible. The narrowest line is the contact between units.

Dashes are useful to indicate special kinds of contacts or structures or very doubtful position, but not to indicate routine doubt. Their drafting adds to the expense and delay in the final map preparation. A line can be mapped solid if it can be correctly positioned to within a few millimeters. Dotted lines, however, are valuable for delimiting buried features. A "scratch" boundary (a contact without an accompanying black line) may separate colored units from blank areas where data are entirely absent.

On cross sections, dashed lines are more useful for indicating doubts about the existence of a unit than about its thickness, which is almost always uncertain. Dashed lines, queried lines, or scratch boundaries may be used, however, where thicknesses are totally speculative or where no further useful inferences can be made.

Lines that are coarser than contacts, usually distinguished by special symbols, are used for structures and for physiographic features, such as crater rim crests. Faults are usually mapped on geologic maps. On maps of some planets, especially those that are not fully understood, the list of line symbols might be quite long. Ganymede, for instance, is mapped with a wide variety of structural symbols representing various furrows, grooves, troughs, rimmed troughs, ridges, and the like (Figure 7.7). In the present stage of mapping these are distinguished, even though future work may show that several are minor variants of a basic type. Each symbol should be fully explained on each map, because symbols may vary among maps.

Like stratigraphic units, structures should be given objective names and their interpretations should be separated from their descriptions in the explanation (e.g., "Scarp: *Interpreted as* fault, locally modified by erosion"). This lesson was learned when better data showed that many lunar "faults" that had been mapped from telescopic photographs and visual observations were, in fact, rather crude alignments of unrelated features. "Lineation" would have been a preferable term. Even "lineations," however, seldom turned out to be significant on the Moon.

Although line symbols are usually in black, they may be colored on special-purpose maps. Features such as rilles (long linear or sinuous depressions) that are considered especially significant may be bounded by contacts and colored as are material units, provided their structural or erosional origin is made clear. They may even be considered as material units. For example, a "rille material" unit may be mapped if it is meant to represent either talus or the exposed, truncated edges of units inferred to be exposed on the rille walls. This was common practice on early lunar geologic maps. Both talus and an exposed section were, in fact, found by the Apollo 15 astronauts in the sinuous rille Rima Hadley. Conventions may be used flexibly to suit special purposes if they are explained and if the distinction between materials and structures is remembered and stated.

A further departure from the Stratigraphic Code may be required by the map scale. As explained in Section 7.2.5, structures may be so dense as to be individually unmappable (e.g., fractured plains material or chaotic material on Mars, grooved material on Ganymede, many units on Venus). The structural characteristic of these units is more apparent than the primary depositional characteristics. In such cases, the unit is mapped on the basis of the structural modification (possibly of more than one true material unit) and this departure is explained verbally.

7.4.6. Explanation

Every geologic map is accompanied by a set of boxes that represent the color and symbol of each map unit and a set of examples of each line symbol. These keys to the map are referred to as the explanation, legend, or key.

The explanation indicates the stratigraphic relations of map units. The most straightforward explanation arranges the boxes in a vertical column, with the box for the youngest unit at the top and the oldest at the bottom. The map units in each system or series are enclosed by brackets labeled with the time-rock names. Boxes of units whose ages are overlapping or uncertain are also shown by brackets, by horizontal positioning, or by diagonal dividing lines if the boxes touch. Boxes for members of a formation or other closely related units touch without a separating space. They may touch at the top and bottom if the relation is one of age. Ends of boxes representing laterally gradational facies touch laterally, commonly with sawtooth boundaries.

Since the early 1970s, terrestrial and planetary geologic maps of the U.S. Geological Survey have employed a more complicated explanation that includes two arrays of colored boxes, each containing the map symbol. One is a vertical array, the "description of map units," which does not necessarily strictly follow stratigraphic order, although it generally has the youngest units or groups of units at the top. The names, characteristics, and interpretations of the units and definitions of new names are written, in telegraphic style, next to each box. The other array, the "correlation of map units," shows all that is known about the stratigraphic relations of the units (Figure 7.9C). The dimensions of this array depend on the number of stratigraphically distinct map units and lateral facies. Related rock units or provinces are commonly grouped in one or both arrays. One Martian geologic map introduced the useful innovation of parallel explanations for depositional and erosional units (Milton, 1975).

The map may be accompanied by a text giving the overall picture of the mapped geology. The text can include such items as geologic setting, provinces, amplifications of the unit interpretations, overall structure, and geologic history.

7.4.7. Mechanics of map assembly

When the mapper is confident of his or her general understanding of an area, mapping can begin on a stable base that will be used for the final map production. I start this process fairly early, because many changes can be made without damaging a properly constructed base map. Such a base is made of transparent plastic (Mylar, Cronaflex, etc.) that is scale-stable (resistant to shrinkage and expansion) and on which the shaded relief or other terrain portrayal is printed on the back – the so-called left reading. Geologic lines are drawn on the front and can be erased without erasing the base even if the lines are drawn in ink. Symbols are placed on the same map and can be pencilled in preliminary stages of mapping. The base portrayal is in brown or some other nonblack color that will not be confused with the geology and that will not reproduce as strongly as the geology when copied by Ozalid, Xerox, or other processes.

The author must label every patch of a unit, even if the final published version will "carry" the identity of some patches only by the color. Most drafters are not geologists and should not have to guess the author's intention about any matter.

At several stages in the mapping, the plastic map is copied onto paper and the paper copy is colored out. Coloring is an essential test of how well the map units and stratigraphic column have been selected and portrayed. This coloring usually results in updating the original stratigraphic and genetic hypotheses. The new viewpoints are incorporated during further mapping. Coloring also always uncovers innumerable inconsistencies and other technical errors that escape even the most careful examination of the uncolored original.

A common error in geologic mapping is to draw junctures of three units incorrectly (Figure 7.19). For readability, the youngest unit should be shown to cut off the contact between the other two on the map as it does in nature. This relation will automatically be drawn correctly if the youngest unit is mapped first. Contacts created by erosion will show the reverse relations: The contacts that bound an old unit exhumed from beneath younger units will cut off the contacts between the younger units. Coloring will help the author identify incorrect truncation relations, which may otherwise escape even the most experienced mapper.

Separate scale-stable manuscript sheets can include additional information

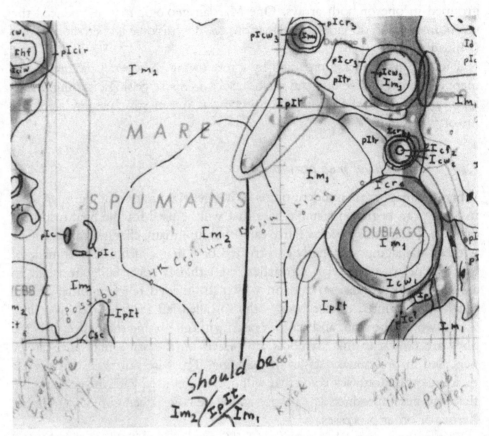

Figure 7.19

Example of map review. Contact-truncation relations and other matters are shown incorrectly on map and are corrected by reviewer in margin. As mapped, unit IpIt (undivided terra material) looks like the youngest unit because it truncates the contact between units Im₁ (older mare) and unit Im₂ (younger mare). The contact should be drawn to suggest diagrammatically a flooding by Im₂ of the two preexisting units.

such as overlay patterns and structures that are to be shown by colored or other special lines. These are registered to the shaded-relief base by registration studs, as are all other stable materials that combine to produce the final map (several parts of the base map, the drafter's scribesheet of the author's linework, "peelcoats" for the colors, etc.). Final maps can be constructed on unstable material such as paper only if they are to be reproduced entirely by photographic or color-scanning processes.

The last colored-out paper copy becomes the "mill copy," which is the guide for drafting and is also used by the author and editors to indicate changes. Authors should color out the mill copy because they will almost always find errors even at this late stage. There is also a colored mill copy of the explanation, including the verbal material, and of a cross section, if one has been constructed. Hand coloring of the explanation boxes and cross section by the author will probably uncover discrepancies between them and the map that have persisted until the end (usually because each change introduces inconsistencies that are not incorporated in all the ancillary working materials). Special supplements to the mill copy may be desirable to provide instructions to the drafter. These might include a sheet to designate line weights clearly and another to clarify a complicated area.

7.4.8. *Consultation and review*

Like all other scientific work, geologic mapping requires more than one person's input. Even experienced mappers consult with others who are working on adjacent areas, comparable areas, or similar problems.

More particularly, maps require reviews by peers. Experience has shown that at least two colleagues should thoroughly examine a map or other publication and frankly express both general and detailed criticisms. A good review requires that the reviewer color out an uncolored copy of the map while examining the data on which it is based. The reviewer should constantly cross-check between map, explanation, and text. Although this task is onerous and time consuming, it always improves a map if done well. The most useful review comments consist of specific criticisms or questions, not a question mark or other vague sign of confusion that will leave the author wondering what is not clear. Comments written in the margins of a map or text, with leaders to the point of difficulty in the body of the map or text, are easier to read and check off than are comments written on the map itself (Figure 7.19). It is in an author's best interest to respond to each comment. Reviewing is also educational to the reviewer and is part of the job of mapping.

Reviews are exploited most productively if the author answers the first before obtaining a second. That is, the author finishes the map three times: once before review, again amid the review process, and again after the last review. This enables two reviewers to view a map as if it were final, not having to go over another reviewer's work. When the reviewers disagree with each other, the author decides which version, if either, to accept.

After review and revision, maps ordinarily undergo a different level of examination, the edit. Editing is the search for mechanical defects in cartographic matters, language, use of geologic nomenclature, and matches with adjacent maps. Much editing is merely the imposition of standards from a style manual. More useful practice is for the editor, normally a nongeologist, to put himself in the place of a naive user of the map and point out things that seem wrong.

Both the reviews and the edits usually uncover matters so familiar to authors that they do not realize the need to state them explicitly. Also, few mappers maintain consistency of presentation over the whole map area and explanation. Different levels of detail or use of different conventions, which may have changed during the course of the mapping, are normally clearer to a reviewer or editor than to the author.

7.4.9. *General guidelines*

The mapper and the reviewer should seek a happy medium between fussiness and carelessness:

1. Small features or fine details should not obscure the big picture, but neither should excessive lumping or "cartooning" gloss over significant differences; worse still, one part of a map should not be constructed in intricate detail while another part is "cartooned."
2. Interpretations should neither uncritically favor a ruling hypothesis nor discuss each conceivable alternative in great detail.
3. Maps should neither uncritically copy others nor strive for originality to the extent of losing sight of the obvious.
4. The list of the defining characteristics of units in the explanation should neither be too brief or otherwise inadequate to allow reproducible mapping nor so excessively complete as to obscure the salient points.
5. The explanation (which is a dictionary) and the prose text should be neither redundant nor inconsistent.

The following additional problems appear repeatedly on manuscripts of geologic maps:

1. Contact lines and structural symbols that are confused because of ambiguous line weight
2. Contacts, especially dashed contacts, that are not connected clearly
3. Unreadable letter symbols
4. Unlabeled patches of units
5. Overprints of symbols and lines
6. Indistinct leaders (the short lines from the letter symbol to a patch of a unit)
7. Inconsistency of symbols between map and explanation
8. Unexplained queries
9. Interpretations based on properties not mentioned among the list of characteristics
10. The reason for age assignments unstated
11. Ambiguous layout of the explanation
12. Incomplete marginal information (scale, credits, source of base, source of data, etc.)
13. Failure to explain the geology for the novice reader
14. Incompleteness (in the hope the reviewer will find the problems)
15. Inconsistencies between the map and the cross section.

These long lists should not inhibit a novice mapper. Their precepts will become automatic with time. However, their observance from the early stages of mapping will help assure a smooth flow of the manuscript map through the review and editing mill. They can be used as checklists. Their purposes are to facilitate map production and to maximize the value of the map both for the mapper and for the user.

7.5. POSTSCRIPT

Geologic mapping can reap considerable rewards. With a relatively minor expenditure of time, effort, and funds, it provided the stratigraphic framework for selection of optimum sites for the Apollo landings and, later, for extrapolation of the findings from these spots to the rest of the Moon. Both the predicted and unexpected results illuminated a range of previously mysterious phenomena and led to new ways of looking at the Moon and other planets and to new paradigms to be tested. I am sure that if geologic methodology had not been available, Apollo exploration would have been content with learning "the" composition and age of the Moon from one or two points. The other planets would have merited only flyby reconnaissance for obtaining astronomical and geophysical data. Imaging of the planets would have been considered scientifically useless. Fortunately, a brilliant technology has been dedicated to the acquisition of spectacular and informative images from as far away as Uranus and Neptune. Geologic mapping is determining the basic architecture and history of planets from Mercury to the far reaches of the solar system and preparing the way for visits to their surfaces.

7.6. REFERENCES

Albritton, C. C., ed. 1963. *The Fabric of Geology*. Reading, Mass.: Addison-Wesley.

Baldwin, R. B. 1949. *The Face of the Moon*. Chicago: University of Chicago Press.

Baldwin, R. B. 1963. *The Measure of the Moon*. Chicago: University of Chicago Press.

Basaltic Volcanism Study Project. 1981. *Basaltic Volcanism on the Terrestrial Planets*. Houston: Lunar and Planetary Institute; New York: Pergamon Press.

Carr, M. H. 1975. *Geologic Map of the Tharsis Quadrangle of Mars*. U.S. Geological Survey Map I–893.

Carr, M. H. 1981. *The Surface of Mars*. New Haven: Yale University Press.

Carr, M. H., ed. 1984. *The Geology of the Terrestrial Planets*. NASA SP–469.

Darwin, C. R. 1958. *Autobiography,* with original omissions restored. N. Barlow, ed. New York: Harcourt, Brace.

Gilluly, James, Waters, A. C., and Woodford, A. O. 1951. *Principles of Geology*. San Francisco: W. H. Freeman.

Greeley, R., and Carr, M. H., eds. 1976. *A Geological Basis for the Exploration of the Planets*. NASA SP–417.

Guest, J. E., Bianchi, R., and Greeley, R. 1988. *Geologic Map of the Uruk Sulcus Quadrangle of Ganymede*. U.S. Geological Survey Map I–1934.

Guest, J. E., and Greeley, R. 1977. *Geology on the Moon*. London: Wykeham.

Masursky, Harold, Colton, G. W., and El-Baz, F., eds. 1978. *Apollo over the Moon: A View from Orbit*. NASA SP–362.

McCauley, J. F. 1967. The nature of the lunar surface as determined by systematic geologic mapping. In Runcorn, S. K., ed., *Mantles of the Earth and Terrestrial Planets*. New York: Interscience, pp. 431–60.

McCauley, J. F., Guest, J. E., Schaber, G. G., et al. 1981. Stratigraphy of the Caloris basin, Mercury. *Icarus* 47: 184–202.

Milton, D. J. 1975. *Geologic Map of the Lunae Palus Quadrangle of Mars.* U.S. Geological Survey Map I–894.

Morrison, David, ed. 1982. *Satellites of Jupiter.* Tucson: University of Arizona Press.

Mutch, T. A. 1970. *Geology of the Moon – A Stratigraphic View.* Princeton, N.J.: Princeton University Press.

North American Commission on Stratigraphic Nomenclature. 1983. North American stratigraphic code. *American Association of Petroleum Geologists Bulletin* 67: 841–75.

Offield, T. W. 1972. *Geologic Map of the Flamsteed K Region of the Moon.* U.S. Geological Survey Map I–626.

Plescia, J. B. 1983. The geology of Dione. *Icarus* 56: 255–77.

Pohn, H. A., and Offield, T. W. 1970. Lunar crater morphology and relative age determination of lunar geologic units – Part I. Classification. In *Geological Survey Research 1970.* U.S. Geological Survey Professional Paper 700-C, pp. 153–62.

Scott, D. H., and Carr, M. H. 1978. *Geologic Map of Mars.* U.S. Geological Survey Map I–1083.

Scott, D. H., McCauley, J. F., and West, M. N. 1977. *Geologic Map of the West Side of the Moon.* U.S. Geological Survey Map I–1034.

Shoemaker, E. M. 1962a. Exploration of the Moon's surface. *American Scientist* 50: 99–130.

Shoemaker, E. M. 1962b. Interpretation of lunar craters. In Z. Kopal, ed., *Physics and Astronomy of the Moon,* pp. 283–359. New York: Academic Press.

Shoemaker, E. M., and Hackman, R. J. 1962. Stratigraphic basis for a lunar time scale. In Z. Kopal and Z. K. Mikhailov, eds., *The Moon,* pp. 289–300. London: Academic Press.

Smith, B. A., et al. 1982. A new look at the Saturn system: The Voyager 2 images. *Science* 215: 504–537.

Spudis, P. D. 1985. A mercurian chronostratigraphic classification. In *Reports of Planetary Geology and Geophysics Program – 1984.* NASA Technical Memorandum 87563, pp. 595–7.

Stuart-Alexander, D. E., and Wilhelms, D. E. 1975. The Nectarian System, a new lunar time-stratigraphic unit. *U.S. Geological Survey Journal of Research* 3: 53–8.

Trask, N. J. 1971. Geologic comparison of mare materials in the lunar equatorial belt, including Apollo 11 and Apollo 12 landing sites. In *Geological Survey Research 1971.* U.S. Geological Survey Professional Paper 750-D, pp. 138–44.

Varnes, D. J. 1974. *The Logic of Geologic Maps, with Reference to Their Interpretation for Engineering Purposes.* U.S. Geological Survey Professional Paper 837.

Wilhelms, D. E. 1970. *Summary of Lunar Stratigraphy – Telescopic Observations.* U.S. Geological Survey Professional Paper 599-F.

Wilhelms, D. E. 1972. *Geologic Mapping of the Second Planet.* U.S. Geological Survey Interagency Report, Astrogeology 55.

Wilhelms, D. E. 1980. *Stratigraphy of Part of the Lunar Near Side.* U.S. Geological Survey Professional Paper 1046-A.

Wilhelms, D. E. 1987. *The Geologic History of the Moon.* U.S. Geological Survey Professional Paper 1348.

Wilhelms, D. E., and McCauley, J. F. 1971. *Geologic Map of the Near Side of the Moon.* U.S. Geological Survey Map I–703.

APPENDIX I

Map formats and projections used in planetary cartography

RAYMOND M. BATSON

A-I.1. FORMATS

A variety of map scales and formats are used to make maps of the planets. This appendix describes and illustrates both those currently in use and those planned for future mapping programs (Tables A-I.a through A-I.h, Figures A-I.1 through A-I.12).

The planet Mars is treated individually in the tables because it is the only solid-surfaced planet besides Earth for which polar flattening significantly affects map projections. A value of 1/192 (eccentricity = 0.101929) has been adopted for computing aereographic (Mars geographic) latitudes for Mars map projections (deVaucouleurs and others, 1973). Geocentric (spherical) coordinate systems are used for all other objects, including very small, irregularly shaped bodies like the Mars moon, Phobos.

A-I.2. SHEET DESIGNATIONS

A rather bewildering assortment of designations are assigned to each planetary map sheet, each of which serves some specific purpose or user. Maps published by the U.S. Geological Survey under the "Miscellaneous Investigations" series (which includes both planetary and terrestrial maps) are given serial numbers in the order in which they are submitted for publication. These serial numbers are prefixed by "I-" and must be used when purchasing maps from USGS distribution centers.

Most map sheets or quadrangles (terrestrial and planetary) are traditionally named for conspicuous features lying either completely or largely within their boundaries. Lunar maps were named for topographic features because these could be seen with telescopes. The Mars 1:5,000,000 series was designed before a comprehensive view of Martian topography was available; therefore, the 1:5,000,000 quadrangles were named for classical albedo features visible through Earth-based telescopes. Quadrangles on Mercury were named after both topographic and albedo features mapped by astronomers. The names based on albedo features are shown in parentheses below the names based on topographic features. Map sheets in subsequent series have been named only after conspicuous topographic features have been identified and delineated within their boundaries.

Many technicians and other users become impatient with long Latin or Greek names and prefer to use a simple numbering system. This system consists simply of a letter code identifying the planet and sequential numbers for quadrangles. This is the system used in the quadrangle layouts in Figures A-I.1 through A-I.12.

As the number and complexity of map series increases, it becomes increasingly difficult to describe or to file maps according to the abbreviated and essentially random systems described above. A system implemented in the early 1970s is a code designating planet, scale, location on the planet, and type of map. The code is defined in Table A-I.i.

No single sheet designator is satisfactory to all users. The I- number has no mnemonic value, and the four-factor alphanumeric code is cumbersome for conversational use. All designators are therefore shown on maps published since 1973.

A-I.3.   REFERENCES

de Vaucouleurs, G. D., Davies, M. E., and Sturms, F. M., Jr. 1973. The Mariner 9 aereographic coordinate system. *J. Geophys. Res.* 78 (20): 4395–404.

Schimerman, Lawrence A. 1975. *Lunar Cartographic Dossier*. St. Louis, Mo.: Defense Mapping Agency Aerospace Center.

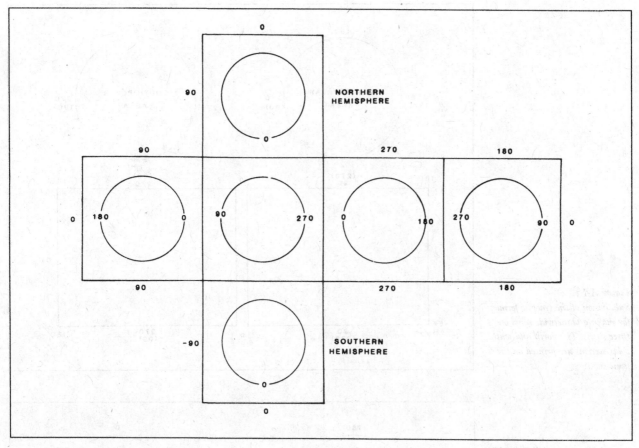

Figure A-I.1
Format for mapping nonspheroidal bodies.

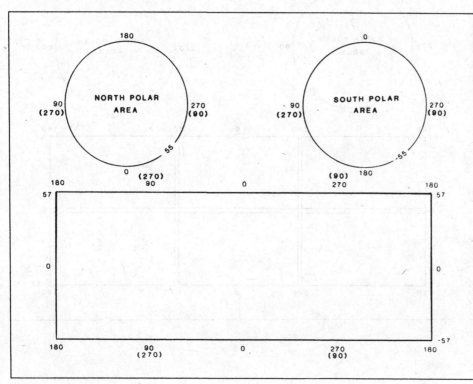

Figure A-I.2
Synoptic format for making planetwide maps on a single sheet.

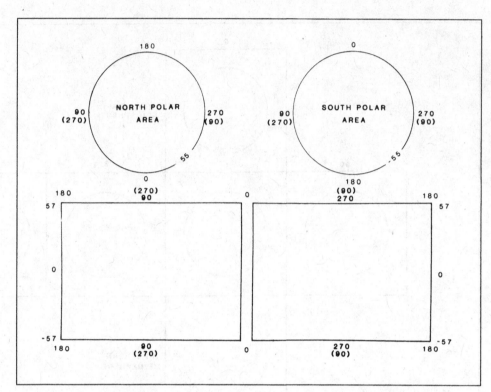

Figure A-I.3
Subdivision of the synoptic format for making planetwide maps on three sheets. The north and south polar sections are printed on the same sheet.

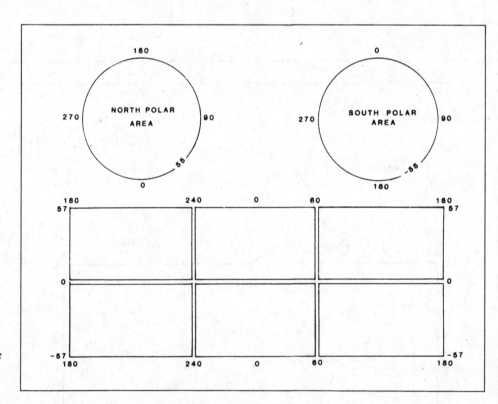

Figure A-I.4
Subdivision of the synoptic format for making planetwide maps of Venus on eight sheets.

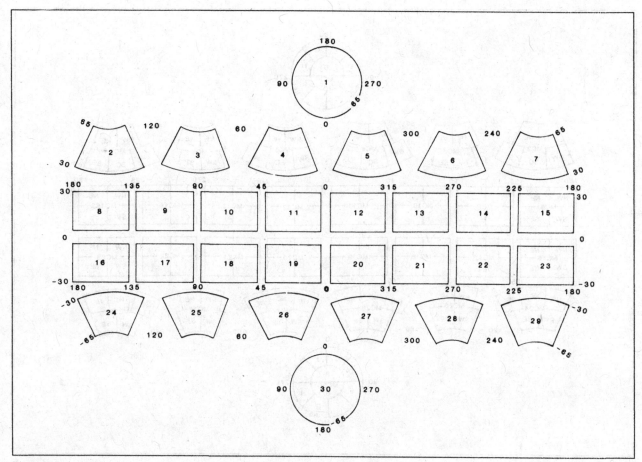

Figure A-I.5
The format for mapping Mars at 1:5,000,000 on thirty quadrangles.

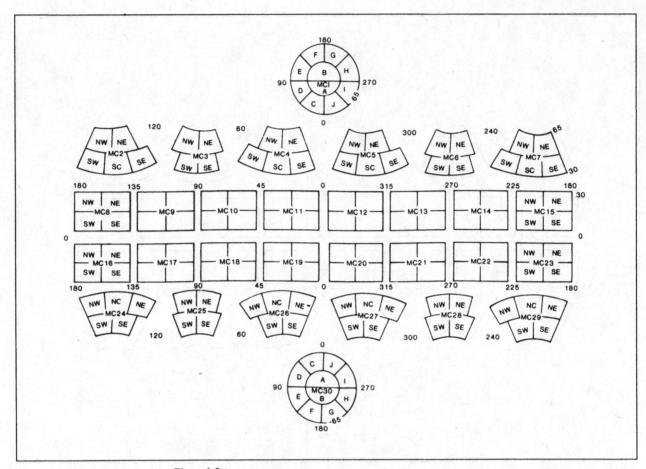

Figure A-I.6
The format for mapping Mars at 1:2,000,000 on 140 quadrangles. This is a subdivision of the 1:5,000,000 mapping format shown in Figure A-I.5.

Figure A-I.7
The format for mapping Mercury-size bodies (Mercury, Ganymede, Callisto, Titan, Triton) at 1:5,000,000 on fifteen quadrangles.

Figure A-I.8
The format for mapping Mercury-size bodies at 1:2,000,000 on eighty quadrangles. This is a subdivision of the 1:5,000,000 format shown in Figure A-I.7.

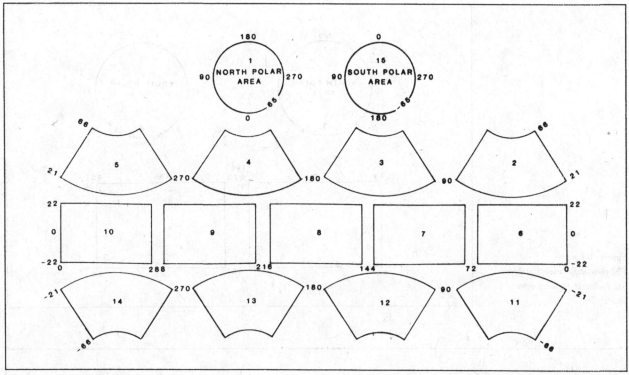

Figure A-I.7 (see facing page)

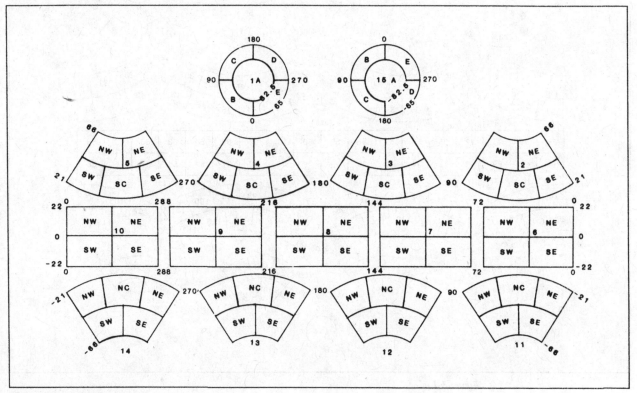

Figure A-I.8 (see facing page)

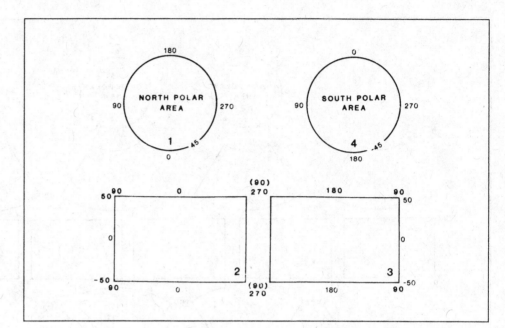

Figure A-I.9
The format for mapping Moon-size bodies at 1:5,000,000 on four sheets.

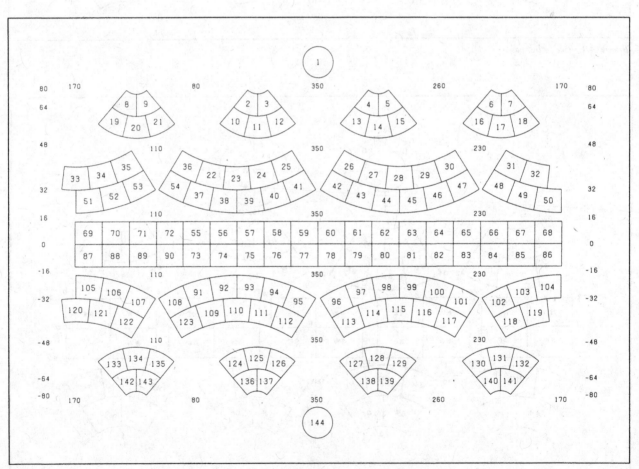

Figure A-I.10
The format for mapping the Moon at 1:1,000,000 on 144 sheets.

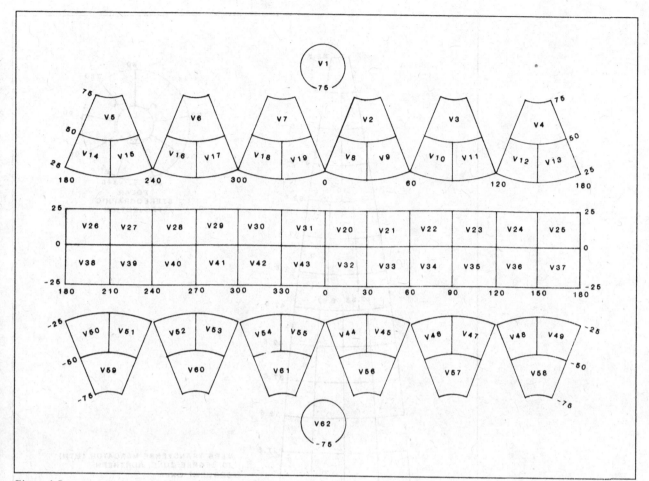

Figure A-I.II
The format for mapping Venus at 1:5,000,000 on sixty-two sheets.

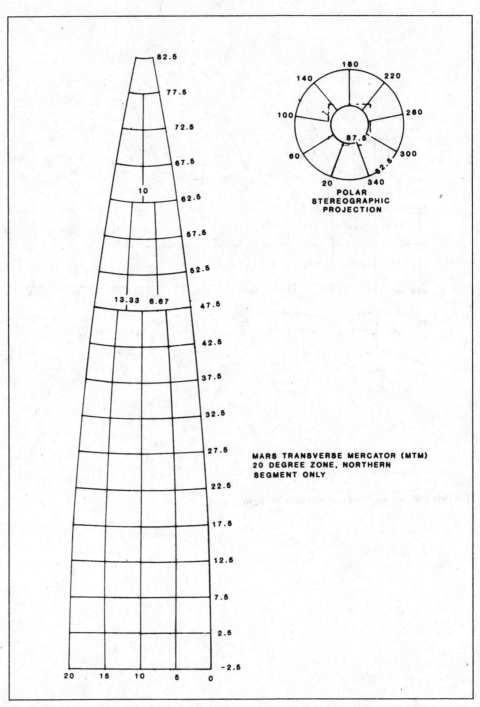

Figure A-I.12
The MTM format for 1:500,000 mapping. It would require 1964 quadrangles to map all of Mars in this format.

Table A-I.a. *Scale factors for synoptic maps.*

Projection	Scale factor	Latitude
Polar Stereographic	1.6354	90°
All spherical planets	1.7883	56°
Mars only	1.8589	90°
	1.9922	60°
Mercator	1.7883	56°
All spherical planets	1.000	Equator
Mars only	1.9922	60°
	1.0000	Equator

Table A-I.b. *Nominal scales used for mapping the planets in the synoptic format.*
The letter M represents million: "1:5M," for example, represents 1:5,000,000.

	Number of sheets required				
	1:50M	1:25M	1:15M	1:5M	1:2M
Mercury	—	—	1	—	—
Venus	1	3	8	—	—
Moon	—	—	—	—	—
Mars	—	1	3	—	—
Io	—	1	1	—	—
Europa	—	1	1	—	—
Ganymede	—	1	1	—	—
Callisto	—	1	1	—	—
Mimas	—	—	—	—	1
Enceladus	—	—	—	—	1
Tethys	—	—	—	1	1
Dione	—	—	—	1	1
Rhea	—	—	—	1	3
Iapetus	—	—	—	1	—
Titan	—	—	—	—	—
Miranda	—	—	—	—	1
Ariel	—	—	—	—	1
Umbriel	—	—	—	1	—
Titania	—	—	—	1	—
Oberon	—	—	—	1	—
Miranda	—	—	—	—	1
Triton	—	1	1	—	—
Nereid	—	—	—	—	—

Dashes indicate that mapping was not done or is not appropriate in the synoptic format at the indicated scales.

Table A-I.c. *Scale factors for Mars quadrangles.*

The equator is the latitude of true scale for the Mars 1:5,000,000 series (Figure A-I.5). The 1:2,000,000 series (Figure A-I.6) was designed with true scale at the standard parallels of the Lambert conics. The enlargement factor required to match a 1:5,000,000 sheet to a mosaic of 1:2,000,000 sheets is therefore 2.221×, rather than 5/2× (2.5×).

Projection	Latitude	Scale factor 1:5,000,000	Scale factor 1:2,000,000
Polar stereographic	90.00	1.1067	0.98300
	65.00	1.1611	1.0313
Lambert Conformal Conic	65.00	1.1611	1.0313
	59.17	1.1259	1.0000
	35.83	1.1259	1.0000
	30.00	1.1532	1.0243
Mercator	30.00	1.1532	1.0243
	27.476	1.1256	1.0000
	0.00	1.0000	0.88819

Table A-I.d. *Scale factors for mapping Mercury, Ganymede, and Titan.*

True scale is at the equator for Mercury maps only. Scale is true at the poles, at the standard parallels of the Lambert Conformal Conic projection bands, and at latitudes 13° N and S on the Mercator projection band for the others. Latitude-overlap zones of 5° between map projections used on the Mercury series were reduced to 2° on subsequent series. A potential 1:2,000,000 series, which uses the scale factors in the right-hand column, is diagrammed in Figure A-I.8.

Projection	Latitude	Scale factor (Mercury 1:5M only)	Revised scale factor
Polar Stereographic	90.00	1.0529	1.0000
	67.50	1.0946	N/A
	65.19	N/A	1.0484
Lambert Conformal Conic	67.50	1.0946	N/A
	65.19	N/A	1.0484
	62.00	N/A	1.0000
	58.00	1.0494	N/A
	30.00	N/A	1.0000
	28.00	1.0494	N/A
	22.50	1.0824	N/A
	21.34	N/A	1.0461
Mercator	22.50	1.0824	N/A
	21.34	N/A	1.0461
	13.00	N/A	1.0000
	0.00	1.0000	0.9744

Table A-I.e. *Scale factors for mapping the Moon, Io, Europa, and Triton at 1:5,000,000.*

The Moon-size range does not lend itself to extensive subdivision at 1:5,000,000 because map sheet size would be inconveniently small.

Projection	Scale factor	Latitude
Polar Stereographic	1.0000	90.00
	1.1716	45.00
Mercator	1.1716	45.00
	1.0000	34.06
	0.82839	0.00

Table A-I.f. *Scale factors for lunar 1:1,000,000 quadrangles.*

The format illustrated in Figure A-I.10 was designed in the early 1960s when information was gathered through telescopes for mapping the Moon at 1:1,000,000. It differs from other planetary map series in that two zones of Lambert Conformal Conics are used. Degrees of latitude and longitude were divided into minutes and seconds on early lunar maps but are shown in decimal degrees here for consistency with the current planetary mapping convention shown in the other tables of this appendix.

Projection	Latitude	Scale factor
Polar Stereographic	90.00	1.0000
	80.00	1.0076
Lambert Conformal Conic	80.00	1.0314
	74.67	1.0000
	53.33	1.0000
	48.00	1.0194
Lambert Conformal Conic	48.00	1.0239
	42.67	1.0000
	21.33	1.0000
	16.00	1.0211
Mercator	16.00	1.0211
	11.0124	1.0000
	0.00	0.9816

Table A-I.g. *Scale factors for Venus 1:5,000,000 quadrangles.*

This format contains two rows of Lambert Conformal Conic projections in each hemisphere. Only one set of standard parallels is used for both, so that adjacent conics will join (Figure A-I.11). Cartographers in the Soviet Union have made a partial series of 1:5,000,000 quadrangles in a slightly different format from the one shown here.

Projection	Latitude	Scale factor
Polar stereographic	90.00	1.0000
	75.00	1.01792
Lambert Conformal Conic	75.00	1.01792
	73.00	1.0000
	34.00	1.0000
	25.00	1.06115
Mercator	25.00	1.06115
	15.90	1.0000
	0.00	0.961725

Table A-I.h. *Scale factors for the Mars Transverse Mercator (MTM) 1:500,000 series.*

Viking returned thousands of images that resolve landforms having dimensions of 100 m and smaller. These images contain unique evidence of poorly understood surface processes, the study of which requires maps at the largest scales supported by image data. The MTM system was designed to support these studies. The series consists of Transverse Mercator Projections in zones 20° in longitude by 5° in latitude, containing 112 quadrangles each (Figure A-I.12). The scale is 1:504,000 at the central meridian of each projection zone and 1:496,000 on the edge. Stereographic projections are used for the polar regions.

Projection	Latitude	Scale factor	Projection	Longitude	Scale factor
Polar Stereographic	90.00	0.9993	Transverse Mercator	10.00	0.9960
	87.50	1.0000		30.00	0.9960
	82.50	1.0038		through	0.9960
				350.00	0.9960

Table A-I.i. *Alphanumeric map-sheet designation codes.*

These codes have evolved somewhat and are therefore not fully standardized for all extraterrestrial maps. Lunar maps made during the 1960s and early 1970s, for example, use a different system of codes from those adopted later for the planets. An abbreviated description of these codes is given in Table A-I.j. A more detailed discussion is given by Schimerman (1975).

STANDARD PLANET DESIGNATORS

MERCURY	H (Hermes)	JUPITER	J	SATURN	S	URANUS	U
		Amalthea	Ja	Hyperion	Sh	Miranda	Um
VENUS	V	Io	Ji	Mimas	Sm	Ariel	Ua
		Europa	Je	Enceladus	Se	Umbriel	Uu
MOON	L	Ganymede	Jg	Tethys	Ste	Titania	Ut
		Callisto	Jc	Dione	Sd	Oberon	Uo
MARS	M			Rhea	Sr		
Deimos	Md			Iapetus	Si	NEPTUNE	N
Phobos	Mp			Titan	Sti	Triton	Nt

NONSTANDARD PLANET DESIGNATORS

MC   Mars Chart: applied to 1:5,000,000 Mars quadrangles, followed by sequential numeric designator. An abbreviated form, often substituted for quadrangle name.

MTM   Mars Transverse Mercator: applied to Mars 1:500,000 maps, followed by latitude and longitude of center point to nearest degree, not separated by solidus (/). This is an abbreviated form, analogous to "MC" designators.

SCALE DESIGNATORS

50M	1:50,000,000	5M	1:5,000,000
25M	1:25,000,000	2M	1:2,000,000
15M	1:15,000,000	1M	1:1,000,000
10M	1:10,000,000	500k	1:500,000

EDITION DESIGNATOR

A number indicating the edition of a map is used only on synoptic maps showing an entire planet on a single sheet, according to the format in Figure A-I.2. This number replaces the location designator described below.

LOCATION DESIGNATOR

Latitude/longitude, to nearest degree (fractional, decimal degrees, or minutes and seconds are not used) nearest to the center point of a map. South latitude indicated by minus sign. Longitude is in degrees west of prime meridian for all planets but Venus, the satellites of Uranus, and Triton, where longitude is east because rotations are retrograde.

VERSION DESIGNATORS (always shown immediately after location designator; may or may not be followed by a secondary version designator)

A   Airbrush map. May or may not include shaded relief. Usually indicates shaded relief with overprinted albedo markings.

CM   Controlled photomosaic. May have been compiled manually or in the computer.

G   Geologic map. Always used alone, without secondary designators.

M   Mosaic. May be uncontrolled or semicontrolled.

OM   Orthophotomosaic.

R   Shaded relief. May or may not have been done with the airbrush.

T   Topographic, with nomenclature. May also be used as a secondary designator if contours and names are overprinted on above types of base maps.

Table A-I.i. (*continued*)

SECONDARY VERSION DESIGNATORS (when these are used, they always follow one of the primary designators listed above)

N   Nomenclature only (no contours)

K   Color. May mean that the base mosaic is made from color pictures or that color coding of some sort is used in the map.

EXAMPLES OF MAP DESIGNATIONS

I-no.	Name	Sequential number	Alphanumeric designator	Description
I–970	Mare Australe	MC–30	M 5M −90/0 R, 1976	Mars 1:5,000,000 map of south pole, shaded relief; published in 1976.
I–1261	Tharsis Southwest	MC–9 SW	M 2M 7/124 CM, 1980	Mars 1:2,000,000 map of area centered near lat 7° N, long 124°; controlled photomosaic; published in 1980.
I–1599	Valles Marineris Region (several maps carry this title, but with different numeric designators)	MTM 00072	M500k 0/72 CM, 1984	Mars 1:500,000 map of area centered near lat 0°, long 72°; controlled photomosaic; published in 1984.
I–1649	Galileo Regio	Jg–3	Jg 5M 44/135 AN, 1984, sheet 1 of 2 Jg 5M 44/135 A, 1984, sheet 2 of 2	Ganymede 1:5,000,000 maps of area centered near lat 44° N, 135°; *AN* of sheet 1 indicates airbrush map showing nomenclature; *A* of sheet 2 indicates airbrush map without nomenclature; both maps published in 1984.
I–1324	Altimetric and Shaded Relief Map of Venus		V 50M 6/60 RKT, 1981	Venus 1:50,000,000 map centered near lat 6° N, long 60° E; shaded relief with color-coded elevation zones; contour lines and nomenclature overprinted; published in 1981.
I–1324	Preliminary Pictorial Map of Rhea		SR 10M 2AN, 1982	Saturn, Rhea 1:10,000,000 planetwide map; second edition; airbrush base showing nomenclature; published in 1982.

Table A-I.j. *Lunar map designators.*

LAC	Lunar Astronautical Chart: applied to 1:1,000,000 maps of the Moon made from information obtained through telescopes during the 1960s. Numbered as shown in Figure A-I.10 (e.g., LAC 78). Only the 44 quadrangles on Earth-facing hemisphere were completed.
LM	Lunar Map: applied to 1:1,000,000 maps of the Moon made from spacecraft data. Numbered as in Figure A-I.10 (e.g., LM 61). Only eleven of a possible 144 of these maps were completed.
LSR	Lunar Shaded Relief: applied to the 1:1,000,000 shaded relief base map printed without contour lines, nomenclature, or graticule. Always accompanies LM or LAC and carries the same number.
LTO	Lunar Topographic Orthophotomap: series of maps made at 1:250,000 from Apollo metric camera pictures. Has contour and nomenclature overprint. "LTO" designation is followed by LAC number in which quadrangle lies, followed by *A*, *B*, *C*, or *D*, depending on whether the map lies in the northwest, northeast, southeast, or southwest quadrant of LAC. The numeral 1, 2, 3, or 4 designates the northwest, northeast, southeast, or southwest subquadrant. LTO84B3 covers southeast subquadrant of northeast quadrant of LAC 84.
LO	Lunar Orthophotomap: same as above, but without nomenclature and contour overprint. Sheet location coding is same as that for LTOs.

APPENDIX II

Halftone processes for planetary maps

JAY L. INGE

A-II.1. INTRODUCTION

The primary requirement for planetary maps is the clear and distinct display
of the images that compose these products. Special photographic procedures
are required to prepare halftone copies of image maps because traditional
processes do not always replicate the full dynamic range of tones contained
in the original compilations. Kidwell and McSweeney (1984) discuss techniques used for special satellite-image maps of the Earth. This appendix describes the approaches used to print different series of planetary maps.

A-II.2. THE HALFTONE PROCESS

Successful lithographic reproduction and publication of photomosaic and airbrush versions of planetary maps begins with the conversion of original continuous-tone images into halftone negatives. Original images are transformed
into an array of tiny dots that vary in size. These halftone dots are duplicated
on press plates that print dots of ink that have only one intensity. Continuous
tones are therefore ultimately shown by solid dots of ink; the larger the dots
in a unit of area, the darker the image.

Typically, halftone negatives are made by placing a contact screen over a
sheet of copy film prior to exposure. The contact screen contains an image
of a fine grid of soft-edged lines that vary in opacity. The greatest opacity is
found at the intersections of these lines and the thousands of tiny windows
between grid-line intersections (Figure A-II.1) allow light to expose the copy
film. When the amount of light passing through the grid is small, as from a
shadow in the original image, light exposes the film only at the center of each
window, producing an image of a tiny dot surrounded by clear film. Very
dark shadow details will underexpose the film, and a halftone dot will not be
present. Similarly, highlights in the original image result in exposure of a
larger area of each window, producing large opaque areas that surround small
clear dots. Bright highlights will overexpose and completely blacken the film.

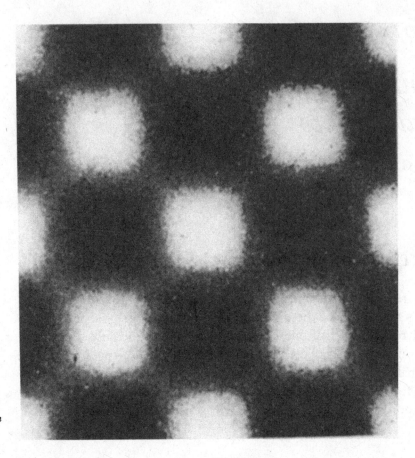

Figure A-II.1
An enlargement of a halftone
screen showing the variation in
opacity of the grid lines.

Figures A-II.2 and A-II.3 illustrate differences in dot size as a function of reflected light from the original image.

An accurate negative halftone copy of a continuous-tone positive image can be made with a contact screen, but the tonal range between highlight and shadow details of the original image (for example, an airbrush drawing) must be similar to the dynamic range of the contact screen. The contact-screen range can be quantitatively determined by use of a reflection-reading densitometer that measures the light reflected by tone steps with known values or by original image shadow and highlight features that nearly underexpose or overexpose the halftone copy. White highlight details have small reflection density values, typically +0.25*, as compared to a sheet of paper, which might have a value of +0.15. Midtone and shadow features usually have reflection values of +0.70 and +1.20, respectively. The density difference (+0.95) between the lightest and darkest features indicates that the density range that can be copied effectively by the contact screen with a single exposure is limited to +0.95; typical contact screens have single exposure ranges of +0.90 to

* These values, when plotted on semilogarithmic paper, show a straight-line relationship when changes in halftone dot area produced by equal changes in continuous tone values are compared.

Shadow Dot
(Negative)

Figure A-II.2
Shadow dots are formed when light passing through the halftone screen exposes only a small area on the film. Only the clear areas will carry ink during printing.

+1.00. Halftones made with these contact screens will contain printing dots that cover about 95 percent of the paper in the darkest shadow details and 5 percent in the highlights.

When the density range of the original copy is considerably wider than the contact-screen range, bright highlight features will overexpose the film and underexposure of dark shadow details will result in clear areas of film. Because the halftone dot defines the appearance of image details, a minimum printable dot will be absent in the highlights and detail will be lost. In addition, various printing problems will be caused by the large areas of solid ink that result from underexposure of an image's dark features. To copy images of this kind, it is necessary to adjust the exposure so that the tonal range or contrast of the original is compressed in the halftone copy. Typically, halftone exposures are made so that the midtone and highlight areas correspond closely to the original, and a minimum printable dot (5 percent) is placed in the lightest features. As a result, many of the darker midtones, in addition to the shadows, can be underexposed, and a large range of densities is represented by clear areas of film. Halftone dots are restored in these areas through the use of a brief "flash" exposure of the copy film and contact screen to a uniform light source. The net effect of this exposure is to add or enlarge very small shadow dots in the halftone negative, but this effect decreases as dots become larger in printing area. For example, two details represented by printable dot percentages of 90 and 96 percent, respectively, might be altered to values of 88

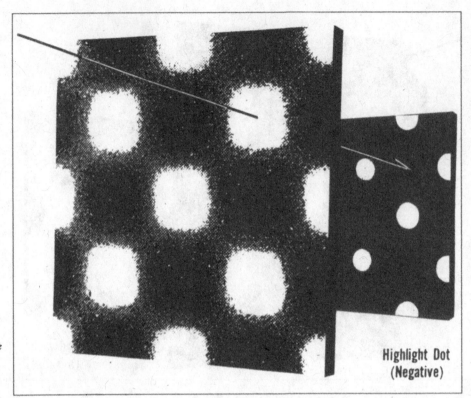

Highlight Dot
(Negative)

Figure A-II.3
Highlight dots are created when large amounts of light passing through the halftone screen expose most of the emulsion of the film. Very little ink will be carried by the small clear dots.

and 90 percent with a flash exposure. Percentages near 60 or 70 percent are mostly unaffected. As a result, the tonal discrimination between various shadow details is compressed and the range between the shadows and mid-tones is subdued in typical halftone reproductions.

A-II.3. MODIFICATIONS APPLIED TO PLANETARY MAPS

To a small extent, exposure methods can be modified to permit less contrast compression of shadow tones in a halftone if some of this compression is also allowed to occur in the lighter features of the halftone. When small highlight details are overexposed and shown with a 0 percent printing dot, and when shadow features with densities greater than 1.30 to 1.40 are underexposed to carry solid ink tints, contrast compressions are distributed between both ranges of an image. Only features lighter than 0.15 are shown without a screen, and details lost as solid tints of ink are limited to the small number of details with densities greater than 1.40. When the highlight density range is overexposed, some midtones are shifted to somewhat lighter values. This shift allows the contact-screen range to provide a greater discrimination between darker tones.

Table A-II.1. *Lithographic processing limitations*.

	Photomosaic density	Conventional screen (%)	Modified screen (%)	Modified halftone approximate ink density[a]
Highlight	0.15	9	0	0.00
Midtone highlight	0.35	25–30	15–22	0.07–0.11
Midtone	0.70	55–60	45–55	0.35–0.40
Dark midtone	0.85	70–75	65–70	0.55–0.70
Light shadow	1.10	9	85–90	0.90–1.00
Shadow	1.30	95	95	1.10–1.25
Dark shadow	1.50	95	Solid	1.25 (solid)
Maximum shadow	1.90	95	Solid	1.25 (solid)

[a]Values taken from reflectance-reading densitometer adjusted to read printing paper as 0.00. Estimates of dot enlargement during printing are approximate.

Ultimately, the actual range of printed versions of these images is limited by the density of available printing inks. Black inks, for example, have a printing density that is no greater than 1.20 to 1.30, a value that is rarely approached by the densities of colored inks. Although printing densities approaching 1.90 can be produced by double-printing black ink as a duotone, there is an additional loss of resolution due to the breakup of detail by another halftone screen. Table A-II.1 is a summary of the lithographic processing limitations imposed on the dynamic range of a typical mosaic of black-and-white images obtained by spacecraft.

Original black-and-white airbrush drawings are reproduced by special halftone methods that result in duotone and tritone color maps. Color is used as a code to identify specific planets. Although two-color combinations are typically used, three ink colors have been used on more recent maps to increase the number of usable color combinations. For example, black and orange inks were used for Mars but black, cyan, and brown inks were selected for Ganymede.

The duotone and tritone process involves making two or more halftone negative copies at different contrast levels and then printing each halftone with a different color ink. These synthetic color separations allow two-color (duotone) and three-color (tritone) reproductions that can range from sheets with colored highlights and neutral shadows to maps with neutral highlights and colored shadows, depending on the color and printing order of the inks. Typical variations in the range of halftones made from a single original image are shown in Figures A-II.4 and A-II.5. The colors of the printing inks are chosen first, because the color determines how much the halftone must be modified to emphasize different parts of a drawing. A pale color, for example, requires larger printing dots than does a bright color to tint image midtones.

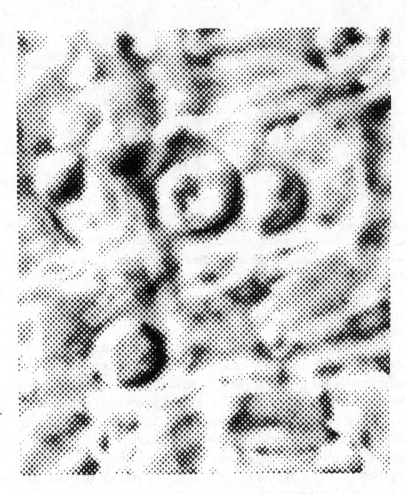

Figure A-II.4
Modification of film exposure creates a plate that contains larger printing dots in the midtones and highlights than the plate in Figure A-II.5.

Dissimilar halftone copies of a single drawing are made when different parts of the contact screen shown in Figure A-II.1 expose the copy film. Because the transition from the clear to the opaque parts of the contact screen provides a gradual but nonlinear suppression of light, increases in exposure do not necessarily produce proportionate increases in the density of the copy film. Increases in exposure tend to compress the density range between highlights and light tones. Normal exposures provide halftone copy with a more linear reproduction of the original image, whereas decreased exposures compress the density range between shadow details. An airbrush drawing's density range, similar to the range of most contact screens, permits selective compression of either the highlight or shadow limit for at least two modified halftones. Table A-II.2 is a comparison of printable dots from two film separations (black and cyan) of the three halftones used in maps of Ganymede. Both of these plates are compressed in the highlight density region to remove printable dots from the highlights. A slightly shorter exposure time of the highlight features allows the black ink to be more dominant than the cyan ink.

The overall light midtone color balance of a map can also be altered by

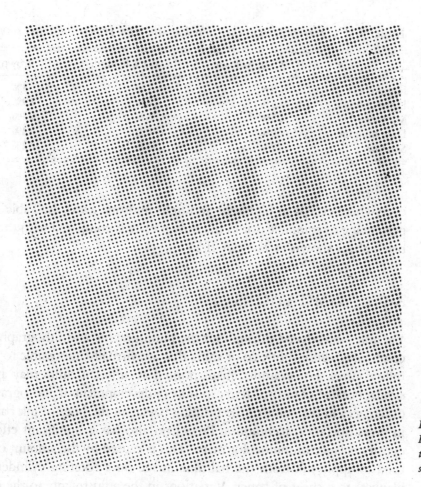

Figure A-II.5
Halftone screen with fewer mid-
tone and highlight dots than
shown in Figure A-II.4.

Table A-II.2. *Screen percentages: black and cyan halftones compared.*

	Cyan Halftone (%)	Black Halftone (%)
Highlight	0	1
Light midtone	5	20
Midtone	25	36
Dark midtone	60	70
Shadow	72	80

modulating the exposure of tint screens. Instead of light reflected from the original drawing being used to modulate the exposure through the contact screen, a uniform light source is used to produce a uniform dot percentage on the copy film. The contact screen is then removed, and the copy film is given a "highlight" exposure with light from the drawing. Although this kind of exposure can fog the unexposed film between the halftone dots, very short exposures allow fogging to affect the highlight detail before interdot fogging becomes a factor. As a result, small percentages of printing dot can be placed

Table A-II.3. *Screen percentages: black, cyan, and brown halftones compared.*

	Cyan halftone	Black halftone	Brown halftone
Highlight	0%	1%	8%
Light midtone	5%	20%	19%
Midtone	25%	36%	23%
Dark midtone	60%	70%	32%
Shadow	72%	80%	32%

everywhere except in highlight and light midtone areas. In Table A-II.3 the percentage of printable dot in the brown plate is compared with the cyan and black plates shown in Table A-II.2.

A-II.4. LIMITATIONS

Variables introduced throughout the photographic and lithographic process combine to cause subtle color shifts that are virtually impossible to predict or fully control. For example, maps printed at different times may appear dissimilar because of inconsistent exposure or development of photographic films, differences in the color of the printing paper, in the hues of ink batches, and in press operation techniques. Press gain also has a significant effect on the density balance of multicolor maps. This is caused by the amount of pressure used to apply ink from the halftone image on a rotating cylinder (transfer blanket) to a sheet of paper. Variations in the adjustments to the circumferences of the rotating cylinders of the plate or transfer blanket, the smoothness of the paper, and blanket resiliency and smoothness are some of the factors affecting how much a dot of ink is squeezed and enlarged by pressure. The greater quantity of ink in a large dot is more readily affected by printing pressure than the smaller quantity in a small dot; this effect distorts tonal balance particularly in the shadows and midtones. A resulting increase or gain of 10 to 20 percent in midtone and shadow screen percentages is not uncommon. Judicious press adjustment can limit this gain to about 5 to 10 percent in the dark midtones and shadows, but there is often a loss of highlight screen. As a result, the same halftone plates can produce maps that vary in appearance when printed on different offset presses or when the operating characteristics of a specific press have changed over time. Additional variables affecting printing conditions can include factors like ink chemistry, plate surfaces, additives in the water, or pressroom humidity. The consistency of appearance of various map series also suffers from the fading of colored inks over a period of time. Although some colors are very light fast (especially black and some cyan inks), colors that use yellow pigments tend to fade when exposed to ultraviolet light.

Although the checkerboard appearance that characterizes many wall-mounted map series is a mute reminder of the limitations of the photographic and lithographic process, controllable standards applied to each stage of processing can produce repeatable results.

A-II.5 REFERENCE

Kidwell, R. D., and McSweeney, J. A. 1984. The art and science of image maps. *Proceedings, American Society of Photogram-* *metry 51st Annual Meeting,* vol. 2, Washington, D.C., March 10–15, 1985, pp. 770–82.

APPENDIX III

Digital planetary cartography

RAYMOND M. BATSON

Table A-III.1 *Scales of digital image mapping. Corresponding map scales are also shown.*

These products are distributed as digital files. Conventional paper reproductions of the digital maps are also published when acceptable analog versions are not available. High-resolution mapping of selected areas is indicated by an asterisk. Mapping to be done from data from future programs (Galileo and Magellan) is indicated by brackets.

| Degrees/ pixel | Index maps | Map scale | | | |
		1:10,000,000– 1:25,000,000	1:5,000,000	1:2,000,000	≥1:500,000
1/16	All	Iapetus Umbriel Titania	Tethys Dione Rhea Ariel	Mimas Enceladus Miranda	
1/32			Europa Io		
1/64			Mercury Ganymede Callisto Triton		
1/128				Io Europa Triton*	Ganymede Callisto
1/256				Mars	Io* Europa*
1/512				Venus	Ganymede* Callisto*
1/1,024					Mars*
1/2,048					Venus* Mars*

Table A-III.2 *Equivalents of digital model pixel sizes, in kilometers per pixel.*
Mean radii are given for highly irregular satellites (Phobos, Deimos, Amalthea, and Hyperion).

Planet	Radius (km)	Digital scale (degrees/pixel)						
		1/16	1/32	1/64	1/128	1/256	1/512	1/1024
Mercury	2,439	2.660		0.665				
Venus	6,052	6.602					0.206	
Mars	3,385	3.692				0.231		0.058
Phobos	11	0.012						
Deimos	6	0.007						
Amalthea	215	0.234						
Io	1,816	1.981	0.990		0.248	0.124		
Europa	1,563	1.705	0.852		0.213	0.107		
Ganymede	2,638	2.878		0.719			0.090	
Callisto	2,410	2.629		0.657			0.082	
Mimas	197	0.215						
Enceldus	251	0.274						
Tethys	524	0.571						
Dione	559	0.610						
Rhea	764	0.833						
Iapetus	724	0.790						
Hyperion	148	0.161						
Miranda	242	0.264						
Ariel	580	0.633						
Umbriel	595	0.649						
Titania	805	0.878						
Oberon	775	0.845						
Triton	1,350	1.473		0.368	0.184			

Table A-III.3. *Scope of systematic digital image mapping of planets.*
Special-purpose maps covering small parts of planets are not shown. Level 2 images often consist of 2000 × 2000 pixel frames with sixteen-bit (two bytes/pixel) encoding. Level 4 mosaics cover 100 percent of each body (except the Uranian satellites, where image data exist only for the southern hemispheres) at indicated resolutions.

Planet	Data type spacecraft	Resolution and volume	Model size		Digital scale (degrees/pixel)
			Meters/pixel	Megabytes	
Venus	Pioneer Venus	Radar altimeter	>10,000	5	1/8
	Magellan	SAR	75	81,500	1/1024
		Radar altimeter	10,000	33	1/16
Mars	Viking Orbiter	Vidicon (4,700 frames)	130–300	42,500	1/256--1/512
Jovian satellites	Voyager	Vidicon (500 frames)	500–5,000	1,000	1/64
	Galileo	CCD (6,000 frames)	100–1,000	12,000	1/128--1/256
Saturnian satellites	Voyager	Vidicon (300 frames)	2000–20,000	600	1/16
Uranian satellites	Voyager	Vidicon (25 frames)	200–20,000	50	1/16
Neptunian satellites	Voyager	Vidicon (25 frames)	200–20,000	50	1/16

Index

Numbers in *italics* refer to figures and tables

Mariner 7, 49, 103, 157
Mariner 8, 49
Mariner 9, 3, 49, 50, 55, 68, 103, 106, 157, 165, 175, 195
Mariner 10, 45, 46, 55, 126, 159, 165
Marius, 115
Marov, M. Ya., 109, 110
Mars, 32, 33–7, 48–51, 61, 62, 70, 93, 164, 165, 208, 241
　airbrush maps of, 77, 78
　Arabia Terra region, 8
　canals on, 35, 37, 145, 146
　chaotic terrain on, 227, 230
　contour map of, 177
　coordinate system of, 147
　crater dating on, 221
　early maps of, 28, 34, 35, 36
　equipotential surface for, 175–6
　erosion on, 231
　geodetic control on, 145–6, 148
　geologic units on, 211
　Hellas, 50
　imaging of, 242
　irregular scarp on, 233
　Map Project, 146
　map scales of, 62, 68, 68
　mapping history of, 54, 55, 56
　"Mariner" Crater on, 48
　Mariner missions, 48, 49
　Nereidum Montes, 50
　nomenclature of, 34, 96, 100, 103–8
　Observer mission, 53, 164, 201
　rotation of, 149
　satellites of, 100, 130 (listed), 208, 225, 226. See also individual satellites
　　control network of, 157–9
　　digital maps of, 82, 83
　　reference surface for, 172
　Sirenum Fossae, 48
　south polar cap, 49
　Soviet missions to, 41, 49, 50
　Task Group for Nomenclature, 109, 110, 111
　topographic control net for, 182
　topographic datum for, 176
　topographic features on, 48
　topographic mapping on, 169, 171, 195–201
　Transverse Mercator System, 69
　Viking mission, 50
Marsden, Brian G., 128
Marth, Albert, 105
Mason, Arnold C., 28
Masursky, Harold, 109, 110
Martynov, D. Ya., 110
Maunder Formation (Moon), 222
Maxwell Gap (Saturn), 128

Maxwell, James Clerk, 125, 126
Maxwell Montes (Venus), 113
Mayer, Simon, See Marius
Mayer, Tobias, 55, 165
　lunar coordinate system of, 142, 147
　lunar map by, 16, 18
Meitner, Lise, 113
Melusine (Ariel), 133
Mercator projection, 32, 34, 37, 46, 60, 61
　characteristics of, 63, 63, 67
　disadvantages of, 63
　scale factor of, 62
Mercury, 32–3, 45, 46, 54, 55, 56, 61, 165, 208
　Caloris Basin on, 46, 224
　control network of, 159, 160
　crater dating on, 221
　geodetic control on, 148
　geologic map of, 46
　Goya Formation, 224
　nomenclature of, 100, 109, 111, 112, 127
　reference surface for, 172
　rotation of, 149
　stratigraphic scheme of, 224
　Task Group for Nomenclature, 109, 110
　Tolstoj Basin on, 224
Meri (Hyperion), 121
Messina Chasmata (Titania), 135
Meteor Crater (Arizona), 234
Metis (Jovian satellite), 130
Meyer, D. L., 144, 144
Millman, Peter, 109, 110
Mills, G. A., 144
Mimas (Saturnian satellite), 52, 129, 130
　nomenclature of, 101, 121, 123, 127
　phototriangulation of, 162
　reference surface for, 172
Minepa (Umbriel), 134
Minnaert function, 88, 89
Minor Planets Names Committee (MPNC), 128, 129
Miranda (Uranian satellite), 53, 130, 131
　craters on, 131
　digital map of, 203
　nomenclature of, 101, 132
　phototriangulation of, 163
　reference surface for, 172
　topographic mapping of, 169, 171, 201
Miyamoto, S., 110
Mommur Chasma (Oberon), 136
Montes Apenninus (Moon), 216, 217, 218, 218, 219

Montes Haemus (Moon), 218, 219, 241
Montes Rook Formation (Moon), 222
Moon, 41, 130, 165, 208
　airbrush maps of, 77
　Alphonsus Crater, 40
　cartographic history of, 12–31
　contour maps of, 174, 192
　control networks for, 156, 157, 157
　coordinate system and, 142, 147
　crater dating on, 221
　early maps of, 12
　equipotential surface for, 174–5
　erosion on, 230
　Euler Crater, 212
　farside, 39
　feature-naming problems on, 102
　Galileo's sketches of, 12
　geodetic control on, 141–5
　geologic maps of, 28, 244
　Gilbert's map of, 13
　Gulf of Cosmonauts, 39
　Hadley Rille, 196, 218, 254
　Harriot's map of, 12, 14
　image diversity of, 242
　Joliot–Curie Crater, 39
　Langrenus' map of, 12
　Lomonosov Crater, 39
　Luna missions to, 38–45
　mapping history of, 54, 55
　map scales of, 62
　Mare Australe, 39
　Mare Crisium, 39
　Mare Fecunditatis, 39
　Mare Humboldt, 39
　Mare Imbrium, 25, 212, 218
　Mare Marginis, 39
　Mare Nubium, 39
　Mare Orientale, 25
　Mare Serenitatis, 25, 97, 218, 219, 241, 247
　Mare Smythii, 39
　Mare Tranquillitatis, 25, 40
　Mare Undarum, 39
　Mare Vaporum, 218
　Montes Apenninus, 216, 217, 218, 218, 219
　Montes Haemus, 218, 219, 241
　Montes Rook Formation, 222
　Mösting A Crater, 142–5, 156, 165
　nearside map of, 244
　Neison's The Moon, 21
　nomenclature of, 97–103
　Oceanus Procellarum, 39
　Orientale Basin, 40
　photographic coverage of, 37
　prephotographic era of, 12–21
　reference surface for, 172

293